Seventy Years
of Garden Machinery

Seventy Years of Garden Machinery

Brian Bell MBE

OLD POND PUBLISHING

IPSWICH

Published by
Old Pond Publishing Ltd
Dencora Business Centre, 36 White House Road
Ipswich IP1 5LT United Kingdom

www.oldpond.com

Distributed in North America by
Diamond Farm Book Publishers

RR3 16385 Telephone Road South
Brighton, ON K0K 1H0 Canada

www.diamondfarm.com

Frontispiece illustration:
Model A Autoculto (Nick Bell)

Cover design and book layout by Liz Whatling
Printed and bound in China

Contents

Acknowledgements

I am indebted to many people but in particular members of the Vintage Horticultural and Garden Machinery Club for their help in compiling this book. Special thanks are due to Garry Butcher, Brian Carter, Bill Castellan, John Catchpole, Stewart Cousins, Stuart Gibbard, John Guest, George Holt, Patrick Knight, Charlie Moore, Ray Smith, Roger Smith and Sandi and Roger Stockham who have provided me with much additional information and new photographs. Thanks are also due to my wife Ivy who managed to find errors in the typescript unnoticed by the computer spell check.

Conversion Table

It is usual to provide conversion tables for those less familiar with the metric system. In keeping with the nostalgic flavour of the book, the information below is offered to help those readers who may be too young to remember the imperial days of pounds, shillings and pence.

1 gallon	=	4.6 litres
1 inch	=	25.4 mm
1 foot	=	300 mm
1 yard	=	910 mm
1 acre	=	0.4 hectare
1cwt	=	50.8 kg
1 ton	=	1,016 kg
1 shilling	=	5 pence
1 pound	=	20 shillings

Introduction

The mechanisation of gardening was already under way in the early 1930s when a wooden wheelbarrow cost 18s 6d and a 12 in cut side-wheel mower could be purchased for £2 12s 3d (£2.61). However, the development of garden machinery in the last seventy years has brought an end to backache and tired legs at the end of a hard day in the garden with a spade or push mower.

This revised and enlarged book traces the changes in powered equipment since the early 1930s for the many and varied gardening tasks including tilling the soil, mowing the lawn and trimming the garden hedge. Petrol engines and electric motors replaced horse, donkey and steam power on the lawns around our stately homes and by the 1940s hand mowers had made life a little easier for the domestic gardener. Few hand mowers are in use today, since we are now able to buy a self propelled or a ride-on rotary mower with an electric starter. The spade and fork still have a place in the garden but even here various shapes and sizes of garden tractor have taken the drudgery out of digging, planting and hoeing the vegetable patch.

There are many anomalies to be found in the world of garden machinery. It is more than likely that some garden tractors will be found with an engine or transmission different to that listed for a particular model. Most of the information concerning engines, gearboxes and other components has been gleaned from catalogues and garden equipment directories. However, there will always be examples of machines with different specifications, usually due to supply difficulties at the time of manufacture or the use of alternative parts during a later restoration or overhaul.

Some machines were sold in their tens of thousands, others were less successful and a few are extremely rare. It is encouraging to know of at least one example of a garden tractor which, thought to be extinct, has been rescued as a result of information given in the previous edition of this book.

Brian Bell
Suffolk, 2007

Chapter I
Two-Wheel Garden Tractors

For centuries, gardeners and smallholders relied on horse-drawn implements and hand tools to prepare seedbeds, kill weeds and harvest their crops, but by the mid-1940s self propelled pedestrian-controlled garden tractors were a common sight. Some vegetable growers used a push hoe such as a one- or two-wheel Planet Junior to sow, hoe and cultivate their crops. However, market gardeners with larger areas under cultivation were often the proud owners of a pedestrian-controlled or walking tractor with a 1 to 6 hp petrol engine.

Two-wheel pedestrian-controlled garden tractors, including the British-made Auto Culto, Monro Monotrac and Duotrac, and the Swiss-built Rototiller, were in use in the early 1930s. By the late 1940s BMB, British Anzani, Garner, Howard, Landmaster, Trusty and other British-made garden tractors, together with the Swiss Simar and a few American machines including the Planet Junior and Gravely were a common sight on smallholdings and in large gardens. There were two distinct groups of garden or walking tractor at the time. The Howard Gem and Landmaster were, for example, classed as self propelled rotary cultivators. Others, ranging from the 1½ hp Farmers' Boy to the 6 hp British Anzani Iron Horse, were suitable for ploughing, cultivating and rowcrop work. Many of the machines in this group could pull a small trailer.

Pedestrian-controlled garden tractors gradually lost favour in the early 1950s as more and more market gardeners and smallholders bought four wheeled Garner, Gunsmith, Newman, Trusty Steed or similar ride-on tractors. Within a few years the more versatile and competitively priced Ferguson TE 20 and other 20 to 30 hp farm tractors had replaced these rather basic four-wheelers. By the late 1980s many smallholders were using Japanese-built four-wheel drive tractors with hydraulic three-point linkage. With the exception of rotary cultivators from Merry Tiller and Dowdeswell, the UK two-wheel garden tractor market was dominated by foreign companies such as Ferrari, Gravely, Honda, Iseki and Kubota.

1.1. Trusty tractors ploughing in convoy. (Roger Smith)

ACRE

The two-wheel Acre rotary cultivator, used to prepare seedbeds, to plough in manure and green crops and to hoe between rowcrops, was made in the 1950s by Acres (Willington) Ltd in Derbyshire. It had a 3 hp four-stroke single-cylinder air-cooled Villiers petrol engine with roller chains running in oil to transmit drive to the wheels and a 15 in wide cultivating rotor.

Dog clutches in the wheel hubs, operated with levers on the handlebars, were used to disengage the drive to the left-hand or right-hand wheel so that the tractor could turn within its own length. The handlebars were adjustable sideways and in height and a tined depth shoe was used to set the cultivating rotor to its maximum working depth of 8 in. In 1951 the Acre rotary cultivator on steel wheels, complete

1.2. The Acre rotary cultivator was made in the early 1950s.

work of two horses as well as carry out various stationary tasks in market gardens, smallholdings and fruit orchards. Following extensive field trials, the first Atco tractors were made in 1935 and it was still in production at the outbreak of World War II. Publicity material for the Atco tractor suggested that the value of this small but efficient machine would be appreciated at once by owners of large estates and growers of every kind of produce. In conclusion it stated that no money, effort or research had been spared in making the Atco tractor the finest of its kind yet produced.

The single-cylinder 350 cc Atco Villiers two-stroke engine, with a cooling fan bolted to the flywheel, developed 8 hp at 2,200 rpm and used about 2½ pints of petrol and oil mixture for an hour's ploughing. A friction-lined cone clutch transmitted engine power through a worm-and-spur

with cultivating tines, an adjustable ridging body, a rake and a toolkit, cost £92 but by the mid-1950s the price had risen to £114. The list of optional equipment included hoe blades and rubber-tyred wheels.

ATCO

Advertisements for the two-wheel Atco garden tractor, made by Charles H Pugh at the Whitworth Works in Birmingham, explained that it could do the

1.3. The first Atco motor cultivators were made in 1935. (Brian Carter)

1.4. The Atco cultivator de luxe was made in the early 1980s.

gearing to provide an independent drive to both wheel axles.

The wheel track was adjustable to any setting from 24 to 34 in by moving the wheels along the axle shafts. The throttle, main clutch, independent wheel clutch controls and a lever to engage the optional power take-off shaft were mounted on the handlebars. The Atco was equipped with a rear towing hitch and either a 3 ft or a 4 ft 6 in wide toolbar for rowcrop work. Attachments included a plough, cultivator tines, hoe blades, two-row ridger, disc harrows, a truck and a 3 or 4 ft cutter bar mower belt driven from the power take-off shaft.

Other equipment made for the Atco garden tractor included a powder duster, sprayer and a three-row Planet-type seeder unit. In 1939 the tractor cost £95 complete with a set of ballast weights and a toolkit. The 3 ft tool frame was £5, the plough with knife and skim coulters was £9 15s 0d and a power take-off shaft was listed at £10.

In 1982 Atco was part of the Birmid Qualcast group. Around this time the standard and de luxe Atco cultivators with their Briggs & Stratton four-stroke engines replaced the Qualcast Super and de luxe Cultimatic garden cultivators (page 69). The 3 hp Atco Cultivator and 4 hp Atco Cultivator de luxe had a forward and reverse vee-belt drive and a secondary chain drive to the cultivating rotor. The working depth was controlled with a pair of adjustable skids and two hinged wheels were provided for transport. Accessories included a pair of rubber-tyred driving wheels for the rotor shaft, a small plough and a toolbar with various attachments.

AUTO CULTO

Allen & Simmonds of Reading, which used 'Pistons, Reading' as its telegraphic address, made the first Auto Culto two-wheel garden tractors in 1926. Within twenty years more than 8,000 were in daily use. The Model A Auto Culto had wooden handlebars, cast-iron wheels and a Villiers Midget two-stroke engine, which was brought to life with a starting handle inserted through the right-hand wheel spokes to engage it with the crankshaft starter dog. A later version of this tractor had steel handlebars, a Villiers X11A two-stroke engine and steel spud wheels.

THE ALL BRITISH
AUTO-CULTO
The Small Motor Driven Cultivator.

THIS IMPLEMENT

CULTIVATES, HOES, HARROWS, PLOUGHS, RAKES, ETC.

ITS USE ENSURES RAPID, EFFECTIVE AND ECONOMICAL CULTIVATION.

FULL PARTICULARS FROM THE SOLE MANUFACTURERS,

ALLEN & SIMMONDS (1925) LTD.
READING, ENGLAND.

1.5. Early Auto Culto garden tractors had two-stroke engines and wooden handlebars.

The early 1940s Auto Culto catalogue listed the 1½ hp DX, 2½ hp EX and 3½ hp FX steel-wheeled garden tractors with turbo fan air-cooled Villiers two stroke engines. The forward speed was variable up to 300 feet per minute. The three tractors had the same dog clutch arrangement to engage drive to the wheels, a throttle lever on the handlebars and adjustable track width. Complete with spanners, toolbox and grease gun they cost £55, £63 and £70 respectively. Attachments included a general-purpose plough, a cultivator, disc and spike tooth harrows and a sprayer. Allen & Simmonds also made the Type J Auto Culto Junior, mainly for cultivating between rows of plants and small trees.

In the 1944 Auto Culto catalogue, the Junior was listed at £45. In the same catalogue the company apologised to customers for any delay in delivery, asking them to bear in mind that it was busy executing urgent orders for the Admiralty and Ministry of Supply.

1.6. The Allen & Simmonds catalogue for 1944 explained that the 2½ hp Auto Culto would do the work normally done by two horses.

engine, but in other respects it was similar to the Field Model L. Attachments for both tractors included a plough, tined and rotary cultivators, a sprayer, a saw bench, a potato lifter, an electric hedge trimmer, a grass mower and a trailer. The Model M rotary cultivator was offset so that the operator did not have to walk on the freshly cultivated ground.

Allen & Simmonds garden tractors made at the Thameside works in Reading in the early 1950s included the 7 hp Field GX, the Garden KG model and the Junior with a new 3 hp engine. The GX and KG were supplied with pneumatic-tyred wheels of 20 in diameter suitable for use with the gear-driven rotary cultivator. Larger 22 in diameter wheels for high-clearance rowcrop work were an optional extra. The 22 in wide 3 hp Garden KG model was recommended for use on kitchen gardens, allotments and small nursery plots.

The price of the single-speed Junior with a 3 bhp Villiers two-stroke engine, steel wheels and a 12 in wide toolbar, had risen to £58 10s 0d by 1949 when attachments included a front hoe, a strawberry hoe and a set of discs.

In the late 1940s Auto Culto garden tractors included the Auto Culto Field Model L, the Kitchen Garden Model M, costing £115 and £103 respectively, and the Type J Auto Culto Junior. The single-speed Field Model L had a 2½ hp Villiers Mk 25A two-stroke engine, a dog clutch to engage drive to the pneumatic-tyred wheels, front and rear toolbars, a 130 rpm power take-off and the track setting was adjustable from 22 to 57 in.

The smaller Auto Culto Model M had a 1½ hp Villiers two-stroke

In 1949, an unusual starting device appeared on some Auto Culto tractors. The principle was similar to a kickstart but with a push-handle and a strong return spring attached to the engine cowling. The handle, which could also have been used for a garden fork, was pushed down to turn the crankshaft

1.7. The first Auto Culto Midget garden tractor was made in 1950. (Roger Smith)

and the makers claimed that with the help of the spring it was not difficult to overcome cylinder compression and start the engine.

Designed for the smaller garden the Auto Culto Midget, with an overall width of 10 in, cost £36 on steel wheels and £37 10s 0d with rubber tyres when it was launched in 1950. There was a choice of a 1 or 3 hp two-stroke Villiers engine, a lever on the handlebars engaged the dog clutch drive to the wheels and, depending on the throttle setting, the Midget did its work at speeds of 1½ to 3 mph. The range of accessories included a set of cultivator tines, hoe blades, ridging body, seeder unit, discs and a 12 in cut lawn mower.

In 1951 Allen & Simmonds celebrated the Festival of Britain with the launch of the Festival Auto Culto de luxe. The Festival, intended for fruit and market garden work and priced at around £130, had a 4 hp Villiers Mk 40 four-stroke petrol or tvo engine, a friction clutch and a two forward and one reverse speed gearbox. Speeds on standard wheels varied from 1¼ to 3 mph. The wheels were adjustable for rowcrop work and power turning was provided by independent clutches in the wheel hubs. Implements for the Auto Culto Festival which were also suitable for the Midget, Model L and Model M, included a front toolbar, plough, rotary cultivator, sprayer and a grass mower.

The Auto Culto de luxe, also introduced in the early 1950s, had a 4 hp petrol or tvo four-stroke engine, centrifugal clutch, two forward and one reverse gears and power steering. Attachments for the de luxe model included a plough, sprayer, twin-cylinder compressor and the quick-attach Roticultivator – the Allen & Simmonds version of a rotary cultivator.

The Auto Culto Autogardener was launched at the 1954 Smithfield Show. It cost £75 and had a wide selection of implements, including a plough, hoe blades, a rotary cultivator and a 2 ft 9 in wide cutter bar. Sales literature explained that the aptly named Autogardener had been introduced to supply the missing link for the many people who were no longer able to secure trained labour. The specification included a 1.9 hp Villiers Mk 15 four stroke engine which was started with a hand lever and ran for up to six hours on a gallon of fuel. The Autogardener had a lever-operated plate clutch, a three-speed gearbox with forward speeds of ½ to 1½ mph. The two higher speeds were

recommended for shallow cultivations and the lowest was for deep digging. The handlebars could be offset to avoid walking on cultivated ground.

The cultivating rotor was driven by a three-speed power take off shaft running at 120, 250 or 350 rpm which could be used with a hedge trimmer and an air compressor to spray paint or apply weed killers, pesticides, etc. There were three versions of the Auto Culto Autogardener in 1956. The Mk I with three forward speeds had ratchets in the wheel hubs, the Mk II had fixed wheels and the Mk III, also with ratchets in the wheel hubs and three forward gears had the added luxury of a reverse gear.

Allen & Simmonds products in the late 1950s included the Horti Culto, Auto Culto Universal, Auto Culto 600 and the Autogardener. The Horti Culto rotary cultivator with a 1.2 hp four stroke engine was advertised as the complete home gardener. In 1959 it cost £57 and could be used for hoeing, ploughing, clearing scrub, mowing lawns and trimming hedges.

Introduced in 1958, the Auto Culto Universal had a 3 hp Villiers Mk 15 four-stroke overhead-valve engine, a single-plate clutch, a three forward and one reverse gearbox with a top speed of 2 mph and a three-speed power take-off. The handlebars could be adjusted for height and offset to either side of the machine.

In 1958 H B Holtum & Co of Cambridge made an expanding front toolbar for the Universal which was simple to adjust and saved time when working in crops planted at different row widths. The Auto Culto 600, which cost £175, completed the Allen & Simmonds range in the late 1950s. In 1961, Auto Culto International of Reading introduced the Midgi Culto. This 12 in rotary cultivator had a 150 cc Villiers engine and a three-plate Albion dry clutch used to engage drive to a three forward and one reverse gearbox and a four-speed power take-off. Accessories including an 18 in rotary grass cutter, furrower and a toolbar with cultivator tines and hoe blades were attached to the tractor with bayonet-type connectors. A flexible shaft from the power take-off was used to operate a hedge trimmer, a chainsaw and a small saw bench.

The improved Universal Autogardener superseded the original Autogardener in 1962. The new model had an overhead-valve single-cylinder 3 hp air-cooled Villiers engine, three forward gears and one reverse, a top speed of 3 mph and a four-speed power take-off shaft. An extended range of implements included a potato lifter, a log saw, a pump, a hedge cutter and an inter row hoe on a front or rear toolbar.

In the mid-1960s the Auto Culto garden tractor range included the Universal Autogardener, the Mk IV Autogardener with 3½ hp BSA engine and the Auto Culto Mk IX with a rotary cultivator, a plough, a toolbar and a front-mounted cutter bar. The Mk IX specification included a 5½ hp

1.8. The Midgi Culto was selected for display at the London Design Centre.

Villiers/JAP petrol or two engine, an Albion three-plate dry clutch, three forward speeds and one reverse, chain drives to the rotary cultivator and wheels, and two power take-off shafts.

Introduced in 1963, the Auto Culto 65 rotary cultivator had a 3 hp Villiers four-stroke engine, mounted above a 12 in wide tiller rotor which could be extended to 36 or 48 in. The vee-belt drive to an enclosed roller chain on the tiller rotor shaft was tensioned with a dead man's

1.9. The Auto Culto Universal Autogardener did four to six hours work on a gallon of petrol. (Roger Smith)

1.10. Spanners were not required when attaching implements to the Auto Culto 55.

clutch lever on the handlebars. A pair of wheels attached to the rear of the Auto Culto 65 controlled the working depth and made the machine easier to handle when rotary cultivating. The tiller rotor was replaced with a pair of driving wheels for use with a rotary grass cutter, a rear toolbar with tines or hoe blades and a hedge trimmer or chainsaw driven by a flexible shaft from the power take-off.

In the mid-1960s the Auto Culto 55 and Auto Culto 75 with similar specifications to the earlier 65 were added to the Auto Culto International range of garden tractors. The 55 had a 3 hp Norton Villiers four-stroke engine, later replaced with a 3½ hp four-stroke Aspera engine, to drive its 12 in cultivating rotor extendable to 36 in. A 3½ hp four-stroke BSA engine was used on the Auto Culto 75 and additional rotor blades could be added to extend the standard 12 in rotor to give working widths of up to 48 in.

By 1970 Auto Culto International had been acquired by Jalo Engineering of Longham near Wimborne, Cambridgeshire. In 1971, the new owners, initially trading as Allen & Simmonds (Longham) Ltd, introduced the Super de luxe 65 with a 3½ hp Norton Villiers four-stroke engine. It was similar to the earlier Auto Culto 65 with a standard 12 in cultivating rotor extendable to 48 in and a range of attachments including a toolbar, a rotary grass cutter and a flexible drive for a hedge trimmer and chainsaw.

Jalo Engineering dropped the Allen & Simmonds name within a couple of years but continued to make the 150 cc Villiers-engined Midgi Culto, the Auto Culto 55 with a 3½ hp Norton Villiers engine, the Autogardener with a 3 hp Norton Villiers Mk 15 power unit and the 5½ hp Villiers/JAP-engined Auto Culto Mk IX into the late 1970s

BARFORD

The prototype of the Atom tractor made and demonstrated by Mechanised Horticultural Implements of Hampshire in 1945 bore little resemblance to the single-wheel Barford Atom introduced to the gardening public in 1947. The prototype, which the Hampshire company planned to call the Atom Major, had a 3 hp engine with a single six-inch wide track at the front and two large-diameter disc wheels at the rear. The operator, who was seated on the machine, guided it with a motor car type of steering wheel.

Barford Agricultural of Grantham, Lincolnshire, a member of the Aveling Barford Group, had acquired Mechanised Horticultural Implements by 1947 when the Atom Mechanical Gardener with a rope-start Mk 10 Villiers 98 cc four-stroke air-cooled engine and lightweight tubular handlebars was introduced. Fuel consumption was less than one pint per hour and power was transmitted through an oil-bath reduction gearbox and roller chain to the wheel. Drive was engaged with a dog clutch assisted by a friction clutch to absorb the shock when moving off.

The Atom, with a working speed of 1 mph, was advertised as a multi-purpose mechanical gardener with a single cleated steel wheel easy to replace with a roller pattern wheel for working on grassland. Attachments for the Atom included a toolbar with tines and hoe blades, a ridging body and a disc harrow. The power take-off shaft on one side of the reduction gearbox could be used to drive a front-mounted cutter bar or a trolley-mounted 110 volt generator with an electric hedge trimmer.

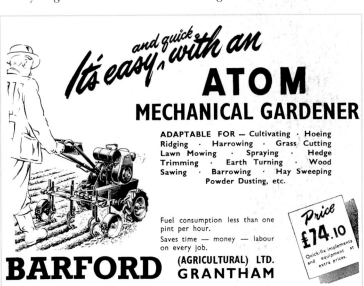

It's easy, and quick, with an

ATOM
MECHANICAL GARDENER

ADAPTABLE FOR — Cultivating · Hoeing Ridging · Harrowing · Grass Cutting Lawn Mowing · Spraying · Hedge Trimming · Earth Turning · Wood Sawing · Barrowing · Hay Sweeping Powder Dusting, etc.

Fuel consumption less than one pint per hour.
Saves time — money — labour on every job.

Price £74.10

Quick-fix implements and equipment at extra prices.

BARFORD (AGRICULTURAL) LTD. GRANTHAM

1.11. The Barford Atom Mechanical Gardener was exhibited at the 1949 Royal Agricultural Show.

Within a couple of years Barfords made an improved Atom tractor with a 1¼ hp air-cooled Villiers engine. The same reduction gearbox and roller chain drive was used, but an improved power take off shaft had a speed range of 300 to 600 rpm. There were two versions of the Atom mechanical gardener. The Atom with twin pneumatic-tyred wheels situated side by side under the engine cost £62 10s 0d in 1950 and the two-wheel rowcrop model with a wheel on either side of the tractor was £69 10s 0d. A two-wheel conversion kit to convert the Atom from single to rowcrop mode was priced at £9 10s 0d.

In 1951, the Barford Atom 15 with a 1½ hp Villiers engine, using no more than a pint of petrol an hour, replaced the Atom Mechanical Gardener. It had a vee belt drive to an enclosed two-speed reduction gearbox with top speeds of 1½ and 3 mph. Barford Agricultural described the two-wheeled Atom 15 as its latest engineering triumph, designed to facilitate and speed up the laborious gardening tasks that confronted the average gardener and nurseryman.

The specification included an adjustable wheel track, differential action on each wheel, a fingertip throttle, belt-tensioning clutch control and quick-release front and rear tool holders. Attachments included a front cutter bar, a cylinder mower, a full range of tillage tools, a snowplough and a hedge trimmer operated by a 110 volt generator vee-belt driven from the tractor power take-off.

In 1953 Barford (Agricultural) Ltd moved to Belton. At this time the Barford Atom 15 cost £84 10s 0d and twenty attachments were included in the price list. It was advertised as the original multi-purpose tractor, miniature in size but mighty in performance.

The 1959 Chelsea Flower Show provided the gardening public with its first sight of the Barford Atom 30 which was advertised by Barfords of Belton as a multi-purpose tractor suitable for ploughing, cultivating, harrowing, ridging and hoeing, cutting

1.12. The Barford Atom 15 made its debut in 1951. (Roger Smith)

1.13. The Barford Atom 30 was launched at the 1959 Chelsea Flower Show.

grass, trimming hedges and haulage work. The two engine options were either a 3 hp Villiers Mk 15HS or a 4½ hp Clinton. A vee-belt and two-speed gearbox transmitted power to the wheels and drive to the rotary cultivator was engaged with a lever-operated vee belt jockey pulley.

In 1960 Barfords of Belton advertised the Atom 30 at £99. The Atom 15 with a 1.9 hp Mk 12 Villiers engine cost £93.10s. An 18 in wide rotary cultivator was added to the list of attachments for the Atom 30 in 1963 when the price of the 3 hp model had risen to £102.

BEEMAN

The Beeman Garden Tractor Co of Minneapolis made two-wheel garden tractors in America from 1914 until the late 1920s. Bourn Engineering Co, based at Grosvenor Place in London, sold the Beeman garden tractor in the UK during the early

1920s, advertising it as a one-horse tractor that would plough, cultivate, harrow and mow at half the cost of owning a horse. The single-cylinder thermo-syphon water-cooled petrol engine developed 1½ hp at the drawbar and 4 hp at the belt pulley. An advertisement suggested that when the tractor was not being used for field work the belt pulley could be used to drive barn machinery, a cream separator, a pump and even a washing machine.

A mid-1940s American tractor directory included an entry for the New Beeman Tractor Co at Minneapolis which listed the

1.14. The Beeman is a typical example of mid-1940s American pedestrian-controlled garden tractors.

Beeman pedestrian-controlled garden tractor rated at 4 belt hp and 2 hp at the drawbar. The specification included a single-cylinder water-cooled petrol engine with a Wico magneto, a Champion ½ in pipe thread spark plug, a starting crank in the flywheel and a double-disc cone clutch. Depending on throttle setting the tractor had a 1 to 3 mph forward speed and the belt pulley ran at 230 to 1,500 rpm

BERTOLINI

Ugo Bertolini built his first two-wheel motor cultivators and scythes at Reggio Emilia in Italy in the early 1930s. Various models made over the years included the mid-1970s two-wheel Bertolini 310 S with the option of a 13 hp petrol engine or a 14, 16 or 18 hp diesel engine and a six forward and two reverse gearbox. Factory-made kits were available for converting the two-wheel Bertolini 310 S into a ride-on four wheeler with equal-sized

1.15. The Bertolini 310S in four-wheel mode with a power take-off driven trailer axle.

wheels and centre-pivot steering or to a long wheelbase 4 x 4 transporter.

In the mid-1980s chainsaw manufacturer Danarm of Stroud in Gloucestershire imported four models of Bertolini garden tractor. There was a choice of a 9 or 10 hp petrol or a 10 hp diesel engine for the Bertolini 306, all with a three forward and one reverse gearbox. In the early 1990s attachments for the 306 included a 26 in wide rotary cultivator, rotary and cutter bar mowers, a plough, a ridger, a saw bench and a sprayer. A similar range of attachments was available for the Bertolini 315 D with a 14 hp diesel engine and a four forward and two reverse gearbox.

In the early 1990s Stratford Power Garden Machinery of Stratford-on-Avon was the UK importer of Bertolini petrol- and diesel-engined garden tractors. There were three models, the 8 hp Bertolini 406 had three forward gears and one reverse while the 9 hp Bertolini 314 and 12 hp Bertolini 318 both had a four forward and one reverse gearbox and a two-speed power take-off.

A rotary cultivator, reversible and right-handed ploughs, cutter bar and rotary mowers, a tined cultivator and a ridger were among the list of the attachments for the Bertolini 406, 314 and 318 two-wheel tractors

BMB

British Motor Boats Ltd, a London-based company, imported small marine engines and Simplicity two

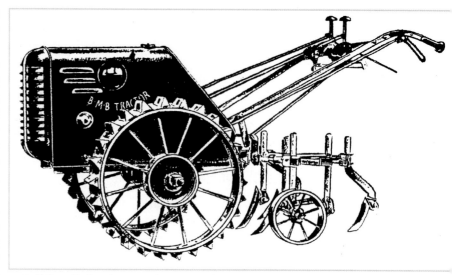

1.16. Shillans Engineering at Banbury was making the BMB light tractor in 1945.

1.17. The mid-1940s Simplicity Model 'C' garden tractor from America was sold in the UK as the BMB light tractor.

wheel garden tractors from America in the 1930s. They were sold in Great Britain as BMB Plow-Mate, Cult-Mate and Hoe-Mate light tractors. British Motor Boats moved to Banbury at the outbreak of war and within a short while a shortage of supplies from America led to limited production of BMB light tractors by Shillans Engineering of Banbury.

In the mid-1940s Brockhouse Engineering bought BMB. It transferred tractor production to Crossens in Lancashire in 1947 and introduced improved versions of the Plow-Mate, Cult-Mate and Hoe-Mate in the same year. Production of pedestrian-controlled BMB garden tractors continued at Crossens until the mid-1950s.

The Plow-Mate, the largest of the three BMB walking tractors, had a 6 hp JAP air-cooled four-stroke engine, a gearbox with two forward and one reverse gears, a power take off and independent brakes. A 6 hp Briggs & Stratton model ZZ four stroke engine was fitted to later models of the Plow-Mate. The flat belt drive from the engine to the gearbox was engaged with an idler pulley operated by an over centre lever on the handlebars. Steel wheels with spade lugs were standard, with pneumatic tyres available at extra cost and the wheel track could be adjusted for rowcrop work.

1.18. The BMB Plow-Mate.

The first British-built Cult-Mate tractors made by Shillans Engineering had a 3 hp engine but this was soon replaced with a BSA 3½ hp power unit. Forward speed was between ½ and 2½ mph and the flat belt drive to the gearbox was engaged with the same over-centre mechanism as the Plow-Mate. Over run ratchets in both wheel hubs provided differential action for turning on headlands.

In order to help the operator work close to plants without damaging them, a toolbar which had two castors and a separate steering lever between the handlebars was designed for inter-row work. Sales literature suggested

1.19. The BMB Cult-Mate had a top speed of 2½ mph.

the Cult-Mate was so easy to handle that a child could manage it and claimed that no operator could fail to appreciate its manoeuvrability or the speed and ease with which it performed many rowcrop operations, bringing to an end back-breaking work and daily fatigue.

Early BMB Hoe-Mate tractors had a two stroke air-cooled 1 hp Brockhouse Spryt engine but by 1948 a

1¾ hp BMB engine provided the power. A lever on the handlebars tensioned the vee belt drive from the engine and differential action was provided by over run ratchets in the wheel hubs. The Hoe-Mate was mainly used for cultivating, hoeing, discing and ridging while a 32 in front-mounted cutter-bar mower suitable for cutting long grass added to the versatility of this little tractor.

1.20. The Hoe-Mate was the baby of the BMB range.

to the front of the tractor and belt driven from a pulley on the engine shaft.

As an aside, in 1948 BMB ran a series of training courses for its dealer staff and salesmen. Five or six students per week attended each course at BMB's own smallholding at London Colney where they were trained to become competent in handling the tractors so that they would then be able to teach owners how to secure the full benefits of mechanisation.

Farm Facilities at Slough used the BMB Plow-Mate engine and transmission unit in producing the Gunsmith light tractor (page 110) and the steam engine manufacturer Garretts of Leiston built a small number of three-wheel BMB Garrett light tractors. The Plow-Mate engine, transmission unit and steel driving wheels were mounted at the back of the tricycle-wheeled chassis, the single front steel wheel was steered with a handlebar and a hand lever was used to raise and lower the underslung toolbar.

BOLENS

Bolens of Port Washington in Wisconsin, better known as a manufacturer of four-wheel garden and lawn tractors (page 98), made its first two-wheel garden tractors in 1919. By the early 1940s the company was trading as the Bolens Products Co. The Bolens range of two-wheel tractors with four-stroke Briggs & Stratton engines included the 1½ hp Super

Sales literature explained that the Hoe-Mate would work all day on less than a gallon of petrol. A testimonial letter from one owner stated that he had bought a BMB because a girl with her Hoe-Mate could do as much work as three or four men with hand hoes.

In 1948 BMB advertised the 6 hp Plow-Mate on steel wheels for £129 10s 0d, the Cult-Mate also on steel wheels with a 3½ hp BSA engine was £85 and the 1¾ hp Hoe-Mate cost £55 on rubber tyres. Additional equipment available for the Plow-Mate and Cult-Mate in the late 1940s included a seed drill, a potato-raising plough, a one-way plough, a cutter-bar mower and a three-wheel trailer. The mower with a 48 in cutter bar for the Cult-Mate and a 52 in cutter bar for the Plow-Mate was attached

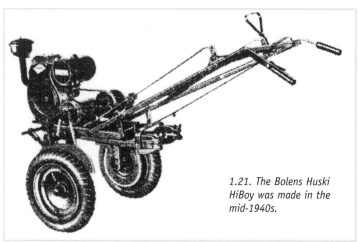

1.21. The Bolens Huski HiBoy was made in the mid-1940s.

Huski, the 2¾ hp Huski Gardener and the 2¾ hp Huski HiBoy. The Super Huski had a vee-belt drive to the pneumatic-tyred wheels. There was a choice of steel or pneumatic-tyred wheels for the Gardener and the HiBoy, both of which had a belt pulley and dog clutches in the wheel hubs controlled with levers on the handlebars to facilitate power turning.

In the mid-1950s two-wheel Bolens garden tractors included the 2½ hp Handi-Ho, the 2 hp Power-Ho and the 4½ hp Gardener, all with Briggs & Stratton single-cylinder engines and one forward gear. A wide selection of Bolens attachments included a plough, cultivator, seed drill, grass mower, sprayer and a dump cart.

Garden Machinery Ltd at Slough imported Bolens 2072 and Bolens 2077 two-wheel garden tractors in the early 1960s. The smaller Bolens 2072 had a 4 hp Kohler engine, a two-speed gearbox and a 16 in wide rotary cultivator. The 7 hp Kohler-powered Bolens 2077 with four forward gears had a top speed of 4 mph.

BRADFORD NU-TRAC

A Mr R Bradford of Hornsey in London designed this unusual single-track pedestrian-controlled crawler tractor in 1948. The air-cooled 825 cc two-stroke petrol or tvo engine ran at 2,500 rpm and used about three gallons of fuel for a full day's work. The air was cleaned and then heated by the exhaust system before entering the cylinder. A roller chain transmitted power from a constant mesh single-speed gearbox to a single vee-shaped track with self-cleaning cleats and a 6 in wide by 18 in long footprint. The clutch and throttle levers were mounted on adjustable handlebars and, depending on the throttle setting, the Nu-Trac was claimed to have a forward speed of 1½ to 4 mph.

A trailed toolbar with a suspension seat and the usual range of tools was an optional extra. Early publicity material explained that Bradford planned to introduce a rubber track for the Nu-Trac for use when mowing and spraying. Front and rear working lights were also to be made available at extra cost.

Light Travel Ltd of Finchley Road in London, which distributed the Bradford Nu-Trac in the early 1950s, advertised the machine as one that could climb a 1:3 gradient and was ideal for working in closely spaced rowcrops.

BRITISH ANZANI

Alessandro Anzani, who was making motor cycles in the late 1800s, established factories in Italy and France to manufacture engines for aeroplanes, speedboats and motor cycles. The British Anzani Engineering Co was formed in 1927 and, after successfully testing a prototype garden tractor in the late 1930s, the first British Anzani Iron Horse tractors were sold in 1940.

Made at Hampton Hill in Middlesex, the Iron Horse had a 6 hp Anzani-JAP four stroke air-cooled petrol engine, an oil bath air cleaner and a Wico magneto using about two gallons of petrol for a full day's work. A centrifugal clutch engaged drive to the wheels through a three forward and one reverse synchromesh Ford gearbox with a top speed of 4½ mph, a worm-and-wheel reduction box and a roller chain final drive.

The centrifugal clutch had a pair of shoes with friction linings

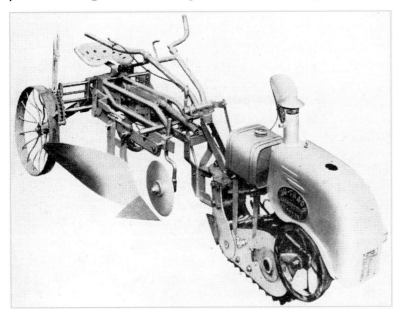

1.22. To do a full day's work the Bradford Nu-Trac used about three gallons of fuel.

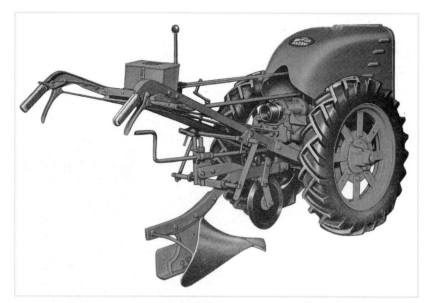

1.23. The belt pulley was at the front of the original British Anzani Iron Horse but it was moved to the back of the gearbox on the post-war version of the tractor.

face of a drum on the output shaft to engage drive to the gearbox. When engine speed dropped below 750 rpm the springs withdrew the shoes from the drum to disconnect drive to the wheels. Independent dog clutches in the wheel hubs, controlled with levers on the handlebars, were used to make very tight turns at the headland. Caution was required when turning as the engine power could swing the handlebars round with some force and potentially knock the operator to the ground.

The 30 in diameter and 4 in wide wheels with steel spade lugs were standard and, although not initially available, pneumatic tyres costing £10 were an optional extra. Track width could be adjusted from 24 to 36 in for

attached to a disc on the engine crankshaft. A spring-loaded linkage allowed the shoes to move outwards against the spring pressure as the crankshaft gained speed and then, when the engine reached a speed of 750 rpm, the shoes came into contact with the inner ploughing and rowcrop work by moving the wheels along on their axles. Optional extras included wheel bands for roadwork, wheel rim extensions and a flat belt pulley at the front of the tractor to drive a saw bench and other stationary equipment.

1.24. There was a choice of steel wheels with spade lugs or welded cleats for the Iron Horse tractor.

1.25. A pair of Rotaped self-laying tracks added £60 to the price of the British Anzani Iron Horse.

The British Anzani Iron Horse cost £107 16s 0d in 1940, although the price had risen to £120 by 1943 and to £130 by 1945. Many wartime Iron Horse tractors left the factory with spade lug wheels and grey paintwork instead of the more familiar orange, perhaps to make them less visible to enemy aircraft

In 1944 Geo Monro of Ingos Works at Waltham Cross demonstrated an Iron Horse fitted with Ingos Rotaped self-laying tracks made by Leeford of London. A press report noted that the Anzani equipped with the Rotaped tracks was able to plough a furrow 14 in wide and 7 in deep on waterlogged land and haul a loaded 15 cwt trailer through a small stream.

A pair of Rotaped tracks, which cost £60, was said to have a working life of 2,000 hours and the use of independent dog clutches in the wheel hubs made it easier to steer the tractor. It was even claimed that an

Iron Horse on Rotaped tracks could be turned within its own axis by using reverse gear and engaging one wheel clutch.

An improved post-war specification in 1945 included the use of the standard 6 hp JAP 5 engine, the top speed was increased to 4½ mph and the belt pulley was re-located on the back of the gearbox. There was also a choice of pneumatic tyres, spade lug wheels or steel wheels with welded cleats which, it was claimed, gave a considerably better grip. Implements for the Iron Horse included a plough, a cultivator, disc harrows, a roll, a seeder unit, a riding seat for use with trailed gang mowers, a cutter bar mower and a trailer.

Hire purchase terms for the British Anzani in 1948 required a £35 deposit with two years to pay the balance. Alternatively, the Iron Horse and a selection of implements could be hired for £3 10s 0d per week.

1.26. Iron Horse tractors at a ploughing match in Surrey. (George Holt)

In 1956 the last of over 10,000 Iron Horse tractors was built but the British Anzani Company remained in business, making ride-on lawn mowers and motor cycle engines. The remaining Iron Horse spare parts were eventually transferred to premises at Aylesford near Maidstone in Kent where the Ridamow and other models of cylinder mower were made. Some of the parts left over from the days of the Iron Horse were used to build a handful of four-wheel ride-on tractors based on the original two-wheeled British Anzani tractor.

British Anzani Engineering at Aylesford also made the 20 in cut Heli-Swift 30 rotary mower for the two-wheel Honda F30 garden tractor. The front-mounted Heli-Swift was driven by two vee-belts. The cutter blade, which was ground along its entire length, had reaper blades at each end.

BULLMAN

In the early 1960s H C Kingcombe (Light Tractors) Ltd of Plymouth marketed the Bullman Major and Bullman Minor two-wheel garden tractors. The main difference was in the engine department: the Major had a 3½ hp four-stroke BSA power unit while a 2 hp BSA engine provided the power for the Bullman Minor. Both had three forward gears with a top speed of 3 mph and a power take-off shaft either running at full engine speed or through a clutch-controlled gearbox at 36, 63 or 100 rpm.

The track width was adjustable from 8¾ to 12¾ in on small wheels and to 22½ in on big wheels. The

Bullman had cultivating widths from 9 to 36 in and the range of attachments included a plough, disc harrows, rotary cultivator, cutter bar, rotary scythe and a potato lifter.

BYRON

In the late 1940s Byron Horticultural Engineering of Hucknall near Nottingham made the blue and red Mk I Landmaster Rotary Hoe. It had a 2½ hp Villiers four-stroke engine, a multi-plate clutch and a three-speed gearbox and was one of the more serious competitors for the Howard Gem Rotavator. Described as the master of every type of land, it cost £89 on steel wheels in 1950 when optional pneumatic-tyred wheels added £5 to the price.

1.27. Other advertisements for the Landmaster rotary hoe explained that it was as strong as a lion yet docile as a lamb.

1.28. The Byron Landmaster rotary hoe was exhibited at the 1953 Royal Smithfield Show.

The Landmaster had a 14 in wide rotor with a maximum working depth of 8 in and the handlebars were adjustable for height and could be swung to either side of the machine. Fitted with a patent retractable stabilising tine to control the working depth, the Landmaster cultivating rotor could be removed in minutes and replaced with a toolbar and used with cultivator tines, hoe blades or a Landmaster seeder unit.

A wider range of attachments was made for the improved 1953 version of the Landmaster Rotary Hoe which cost £100 on pneumatic tyres. These attachments included a plough, cultivator, ridging body, seed drill, sprayer, grass cutter and a 10 cwt capacity trailer. In 1953 Byron Horticultural Engineering changed its name to Landmaster. The Gardenmaster and Landmaster garden tractors (page 54) were made at Hucknall until 1971 when production was moved to Poole in Dorset.

CATCHPOLE

William Catchpole designed the Rowtrac pedestrian-controlled tractor (page 70) in the late 1920s. After employing a local engineering company for a while to manufacture the tractor he sold the patent rights to Geo Monro Ltd in the early 1930s. Catchpole used the proceeds of the sale to design a sugar beet harvester and Catchpole Engineering went on to become a leading manufacturer of sugar beet harvesting machinery.

1.29. Catchpole Engineering Co described the Terra as a multi-purpose power unit.

The company renewed its interest in garden tractors in 1960 when it exhibited the Motostandard Terra horticultural power unit at the Royal Smithfield Show. There were various attachments, claimed to be interchangeable in a matter of seconds so that no time would be lost when switching from rotary cultivating to ploughing, tined cultivating, ridging or rolling.

The Motostandard Terra could also be used for spraying, mowing, sweeping, pumping and hauling a trailer. Sales literature also indicated that when it was not required for land work the power unit could be equipped with a small propeller and used as an outboard motor.

CLIFFORD

Precision engineers Clifford Aero and Auto, which was established at Hall Green in Birmingham in 1912, introduced the Clifford Model A rotary cultivator and the Model B combined cultivator and rotary hoe in 1947. The Model A Mk I had a 5 hp air-cooled JAP engine with a decelerator lever on the handlebars to slow the engine when turning at the headland or standing idle. It had a direct drive to the tiller rotor and a single-speed worm-and-wheel reduction gear to the land wheels was engaged with dog clutches. The petrol tank held one gallon which was said to be sufficient fuel for about 2½ hours' work.

The Mk I with a 16 in wide rotor cost £100 and a 22 in rotor added £5 to the price. An optional pair of larger diameter steel wheels increased the cultivating speed of the Model A Mk I from 1 to 1½ mph.

The Model A Mk II had the same 5 hp JAP engine but it was a more sophisticated machine with a friction clutch operated by a lever on the handlebars and a two-speed gearbox with top speeds of 1½ and 3 mph. A reverse gear and wheel clutches for power turning were optional extras. The standard Model A Mark II tractor with a 24 in wide cultivating rotor cost £137 10s 0d but there was a reduction of £1 15s 0d when supplied with a 16 in wide rotor.

The Mk III with a 22 in wide rotary cultivator priced at £160 was the most expensive version of the Clifford Model A. The specification included a 6 hp side-valve JAP engine with a decelerator control, a lever-operated friction clutch, two forward gears and one reverse, independent wheel clutches for easy headland turns and a worm-and-wheel drive to the land wheels.

The forward gears, selected with a lever on the handlebars, provided speeds of 1 and 2 mph on the standard diameter steel or pneumatic-tyred wheels or

1.30. The Clifford Model A Mk I with a 5 hp JAP engine and a 22 in wide spring tine tiller rotor cost £105 in 1950. (Roger Smith)

1.31. The Clifford Model A Mk III did about two hours' work on a gallon of petrol.

1½ and 3 mph on the optional large diameter wheels. Reverse was engaged with a second lever on the handlebars, which was automatically disengaged when the lever was released.

Optional extras for the Model A included a petrol/tvo conversion kit and a power take-off with a stand used to lift the tractor clear of the ground when it was used to drive a saw bench from a belt pulley on the power take-off shaft. Implements for

1.32. The Clifford Model B cost £125 in 1950.

extensions to increase the track width to 18 in. Implements for the Model B included a 16 in Webb lawn mower, a front cutter bar mower, plough, a generator with an electric hedge trimmer, a 24 in toolbar and an Evers & Wall medium/high pressure sprayer.

The Clifford Mk IV with a BSA 500 cc air-cooled engine introduced in 1953 was the first of a new generation of Clifford rotary cultivators. In 1956 growers requiring extra power were able to specify the alternative 600 cc JAP instead of the 500 cc BSA power unit. Both engines had a dry sump with a force-feed lubrication system, with a pump to circulate oil from a tank under the engine hood to the moving parts. The Clifford Mk IV had two forward gears and one reverse, power steering, a 16 or 22 in wide rotor with rigid or spring tines and handles that could be offset to either side of the machine.

An advertisement at the time explained that the tough and powerful Mk IV could work at a rate of one acre a day with ease when rotary cultivating up to 9 in deep, inter-row hoeing, ridging or mowing. The engine was started with a crank handle which the

Clifford Model A garden tractors included a plough, hoe blades, ridging body, seeder unit, roll, mower, potato spinner, soil shredder and a water pump.

The Clifford AP ploughing unit, AM mowing unit and AS spraying unit had the same general specifications as the Clifford Model A Mk I rotary cultivator. The AP ploughing tractor on steel wheels with a set of front balance weights cost £142 10s 0d. The Clifford AM mower which cost £129 10s 0d could have a 3 ft or a 3 ft 6 in wide front cutter bar with a triple vee-belt drive from the back of the gearbox. The Model AS spraying unit with front-mounted spray bar and a hand lance for spraying fruit trees was £144 10s 0d.

The Clifford Model B inter row hoe and rotary cultivator was designed for greenhouse and rowcrop work. It had a lever-operated friction clutch, two forward speeds, a 12 in wide cultivating rotor and a JAP 5 hp air-cooled engine which did about three hours' work on a gallon of petrol. Optional extras included a reverse gear and wheel hub

1.33. The Clifford Mk IV was introduced in 1953. (Roger Smith)

27

instruction manual explained should be started by pulling the handle upward from the 6 o'clock position. It was stressed in block capitals that the user should never attempt to swing the engine and never push the starting handle down.

The new Clifford Mk I cost £68 10s 0d when it was launched at the 1954 Royal Smithfield Show. It could dig, hoe, ridge, cut grass and trim hedges and used no more than a gallon of fuel to do eight to ten hours' work. It had a 1 hp 80 cc JAP two-stroke engine with a multi-plate clutch running in oil, two forward gears and a power take-off at both ends of the gearbox. The gear-driven 12 in tiller rotor, controlled by a separate clutch, could be attached at the front or rear of the power unit.

1.34. The Clifford Mk I cultivator could be used with a front- or rear-mounted 12 in wide tiller rotor.

In 1955 there was a choice of an 80 cc JAP or a 147 cc Villiers two stroke engine which was recommended for tough working conditions. The Clifford Mk I could be used with the 12 in wide tiller rotor and with either a narrow 6 in rotor or three 6 in rotors spaced at intervals on the toolbar to suit various row widths.

In 1955 a completely re-designed Mk IV Clifford rotary cultivator running on petrol with either a 500 cc BSA or a 600 cc JAP l engine or on diesel with a 500 cc Sachs engine replaced the previous model. The steering column was lowered to allow the machine to pass under low branches and a quick-change system made it a simple task to fit any of the four different types of tine to the 16 or 22 in wide cultivating rotor. The new Clifford Mk IV had a lower bottom gear and a guard to protect the engine from being damaged by a boulder or tree stump.

At the 1955 Smithfield Show Clifford announced the new Mk 0 and Mk II rotary cutivators, an Allman sprayer unit for the Mk I and a new front mower made by H C Webb, also for the Mk I. The smaller Clifford Mk 0 had an 80 cc two-stroke engine, a single-speed gearbox, a multi-plate clutch and a 12 in tiller rotor controlled with a separate clutch. The Mk II cultivator was equipped with a 250 cc two-stroke engine, a two-speed gearbox, a multi-plate clutch and a 16 in wide tiller rotor. Both models had front and rear power take-off shafts.

In 1956, Clifford Aero & Auto changed its name to Clifford Cultivators and moved to premises at West Horndon in Essex where it made the Mk I and Mk IV Clifford rotary cultivators. Following an agreement between Clifford and Rotary Hoes Ltd in 1956 the Clifford rotary cultivation interests were transferred to Rotary Hoes Ltd but the Clifford machines retained their separate identity for a couple of years. The 5 hp Clifford Continental motorised plough with a range of implements including a reversible plough was introduced at the 1957 Smithfield Show. Primarily designed for the export market, the Continental had a four-stroke JAP engine and one forward gear with a top speed of 1½ mph.

The Howard collection at the time consisted of Bantam, Bullfinch, Demon, Yeoman and Gem Rotavators while the Clifford Cultivators catalogue included the Mk I, the Mk IV with optional petrol and diesel engines and the Clifford Super Scythe. Production of Clifford Cultivators had ceased by the time of the 1960 Royal Smithfield Show when the Howard 400, 700, Gem IV, Bantam, Demon, Bullfinch Rotavators and the Hako cultivator were exhibited under the Howard-Clifford banner.

COLEBY

The Rear of Post Office, High Street, Swanley, Kent was the postal address of Coleby Cultivators where Frederick Coleby made his first two-wheel Coleby garden cultivators in the early 1930s. The Coleby Senior and Junior cultivators and the Minor motor hoe remained in production until the mid-1950s, while the two-wheel Coleby Jersey Senior and the four-wheel Coleby Shire tractor were still being made in the late 1950s. The yellow Coleby Senior Jersey Type rowcrop cultivator with red wheels had a 5 to 6 hp four-stroke Coleby engine, a friction plate clutch controlled with a hand lever, a three forward and one reverse gearbox and dog clutches in the wheel hubs providing power steering.

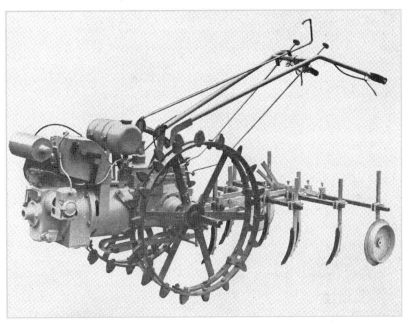

1.35. Only one spanner was required to make all the necessary working adjustments to the Coleby Senior rowcrop cultivator.

The wheel track could be adjusted from 15½ in on single wheel rims to 3 ft when bolt-on axle extensions were used. The handlebars could be offset to either side of the cultivator to allow the user to walk on one side of the machine and it was only necessary to carry one spanner as all the adjusting nuts were the same size. Sales literature informed potential customers that fuel consumption was one gallon per five hours' cultivating and explained that the Coleby was so easy to manoeuvre that it could be handled by either sex with very little tuition.

An improved Senior two-wheel garden tractor was introduced at the 1954 Royal Smithfield Show. It had a 6 hp BSA engine but otherwise the specification was virtually the same as the earlier model. The new model was described as the Jersey Senior in a press report where it was noted that over 350 Coleby machines were working in Jersey and that spares were still being supplied for machines purchased in the early 1930s.

The Coleby Junior rowcrop cultivator was powered by a JAP 4/3 four-stroke 3½ hp engine said to run for seven hours on one gallon of petrol. Forward speed varied between ½ and 1½ mph depending on the throttle setting, and steering clutches assisted turning on the headland.

Standard equipment for the Coleby Senior and Junior included front and rear toolbars with cultivator tines and hoe blades. A plough, a seed drill and a cutter bar attachment were available at extra cost. Coleby Cultivators also made the single speed 1 hp Coleby Minor motor hoe. It was only 7½ in wide, cost £40 in 1951 and was suitable for inter-row cultivating and hoeing.

COLWOOD

In 1945 Dashwood Engineering, based at the Empire Works, Penge in London, introduced the single-wheel Colwood Model A motor hoe. The specification included a 1.3 hp JAP 2A air-cooled four-stroke engine started with a hand lever, an Albion three-speed gearbox with speeds from ½ to 1½ mph and a roller chain drive to the single cleated steel wheel. In 1945, the 13 in wide Model A cost £47 10s 0d complete with a front plant guard, a toolbar with hoe blades and cultivator tines and a toolkit.

The Model A could also be used with a Planet

seeder unit and within a few months Dashwood Engineering had added 14 in cut Shank's New Britisher and Qualcast Panther cylinder lawn mowers to the list of optional attachments. The mower was trailed from the rear toolbar and so it was not possible to use a grass box. Instead a deflector plate was supplied to spread the clippings. A road band was supplied with the mower to prevent the Model A's cleated wheel damaging the lawn.

The single-wheel Colwood Model B Garden tractor with a 1.2 hp JAP 2A or a Villiers Mk 10 four-stroke engine replaced the Model A in 1948. It had a two forward speed Dashwood-Albion gearbox with top speeds of 1½ and 2½ mph and an enclosed roller chain drive to the 14 in diameter steel or optional pneumatic-tyred wheel. Attachments for the Model B included a toolbar with tines and hoe blades, a seeder unit, fertiliser spreader, sprayer, rotary mower and a front-mounted 14 in cut Shanks or New Britisher cylinder lawn mower.

The Goodwood Model M, also introduced by Dashwood Engineering in 1948, was advertised as a new garden tractor designed to take the horse out of horse hoeing. It had a 2½ hp Villiers Mk 25 four-stroke engine started with a hand lever and four forward gears with a speed range of 1 to 2½ mph. Two roller chains and totally enclosed gears transmitted drive to the single 12 in diameter and 6½ in wide steel wheel under the engine. An alternative slightly wider version of the Model M had two 4 in wide steel wheels.

Attachments for the Goodwood Model M included a power take-off shaft driven directly from the engine, a generator and a 16 in diameter rotary grass cutter with four swinging blades driven by a vee-belt from a pulley on the power take-off shaft. The height of cut was adjusted with a small wheel at each side of the rotor hood.

Other accessories included two fully adjustable depth control wheels which could be mounted at the front or rear, a light plough, a ridging body and a 24 in toolbar complete with cultivator tines and hoe blades. A 3 to 4 cwt capacity light truck and a Whitwood seed drill, which could be used with a larger hopper to distribute fertiliser, were also offered.

The Colwood Model RA rotary hoe cost £79 19s 0d

1.36. Depending on availability the Colwood Model B garden tractor was supplied with an air-cooled JAP or Villiers engine.

when it made its debut in 1950. The specification, depending on availability, included a 1.2 hp Villiers Mk 10 or a JAP 2A four stroke engine with a starting lever, a three-speed Dashwood-Albion gearbox and enclosed roller chain drive to the single pneumatic-tyred or optional steel wheel.

Attachments made for the earlier Model B were equally suitable for use with the Colwood Model RA and advertisements suggested that it was an easy task to remove the eight-blade rotor from the tractor and prepare it for toolbar work.

The single-wheel Hornet motorised push hoe with a 98 cc Villiers Midget Mk II two-stroke engine was added to the Colwood garden tractor range in 1951. Launched at the 1953 Royal Smithfield Show, the Colwood Model C was unlike earlier single-wheel Colwood motor hoes. The Model C had two pneumatic-tyred wheels, a 1.9 hp Villiers four-stroke engine, a two-speed gearbox and roller chain drive to both wheels. There were three versions. One had a 12 in wide quick-detach rotary cultivator with a 6 in working depth, another was a basic motor hoe with a tool frame for cultivating and hoeing in rowcrops and the third was a motor scythe with a 3 ft wide front-mounted cutter bar.

In the mid-1950s Dashwood Engineering was acquired by Landmaster. While production of the Colwood RA and Model B Mk I for the export market continued at the Empire Works, the Landmaster Colwood Model B Mk II garden tractor was announced for the home market. It had a 1.3 hp Villiers Mk 12 four stroke petrol engine, a two-speed gearbox and an enclosed roller chain drive to the single wheel.

1.37. The three-speed Colwood Model RA rotary hoe was introduced in 1950.

An advertisement in 1954 explained that the Model B Mk II used no more than four pints of petrol in an average eight-hour day. It was suitable for working in rowcrops, cutting grass and trimming hedges and, unlike the Mk I, it could also be used with a rotary cultivator. The last Colwood garden tractors were made at the Empire Works in the late 1950s.

DOWDESWELL

In 1985 Dowdeswell Engineering bought the Howard Rotavator factory at Harleston in Norfolk from the receiver, along with the production rights for certain models of British-built Howard farm and garden machinery, including the Gem and the 352 rotary cultivators.

Manufacture of the green and orange 352 rotary cultivator, now called the Dowdeswell 352, continued at Harleston but with a 5 hp MAG engine. Dowdeswell also carried on with the Gem but the company was not able to use the Rotavator and Gem names which had been used for Howard rotary cultivators since the days of the Rotehoe Gem 1938. The Gem

became the Dowdeswell 650 pedestrian-controlled rotary cultivator with the choice of a petrol or diesel engine and a 20 or 24 in wide cultivating rotor. The 650 had three forward gears and one reverse, multi-plate transmission and rotor clutches, and a differential that automatically locked when drive was engaged to the cultivating rotor.

Prices bore no comparison to those in the early days of the Gem, starting at £2,559 for the 20 in

1.38. The Howard Gem became the Dowdeswell 650 in 1965.

Dowdeswell 650 with an 11 hp Kohler petrol engine to £3,361 for the 24 in machine powered by a 12 hp Ruggerini diesel engine. A 13 hp Acme diesel engine was fitted to some 20 and 24 in Dowdeswell 650 rotary cultivators. In 1987 the Dowdeswell 650 with a wider 30 in rotor and a 16 hp Kohler petrol engine or a 13 hp Hatz diesel was introduced. The Dowdeswell 352 and 650 rotary cultivators were discontinued in 1991

1.39. A MAG engine was used for the Dowdeswell 352.

EMERY

In 1950, AJ Emery & Son of Halesowen exhibited the Emery Universal rotary cultivator at the Royal Smithfield Show. Power was provided by a 2½ or 3 hp four stroke JAP engine with a dry sump; a pump supplied with oil from a separate reservoir ensured adequate engine lubrication no matter what the angle at which the machine was used. The Universal cultivator had three forward gears with a top speed of 2¼ mph with reverse an optional extra. It had a dry-plate main clutch with secondary clutches to engage drive to the 14 in cultivating rotor and the wheels. An adjustable skid bar was used to set the rotor to its maximum 8 in cultivating depth when working on previously disturbed land.

Sales literature suggested that since the Emery Universal, which cost £85, could be used for ridging, sowing, spraying, grass cutting with a cutter bar mower, hedge trimming and transport work, it would fulfil the needs of the smallholder and market gardener.

Three models of the Emery Two Way Universal Cultivator with different engine outputs and cultivating rotor widths were made at Halesowen in the late 1950s. A 4, 5 or 6 hp air-cooled BSA four-stroke engine was used to drive the 14, 16 or 18 in wide rotors. The three models had a multi-plate main clutch, two forward and two reverse gears, a separate clutch to engage drive to the rotor and a three-speed power take-off. The reversible handlebars, which

1.40. The 3 hp Emery Universal Rotary Cultivator was exhibited at the 1950 Royal Smithfield Show.

enabled the operator to walk in front or behind the machine while using a front- or rear-attached implement, could be offset to either side of the tractor.

Equipment for the Emery Two Way Universal included a plough, ridger and seeder units, cylinder and cutter bar mowers, a hedge trimmer and a saw bench.

FARMERS' BOY

Made by Raven Engineering at Hampton Wick in Surrey the Farmers' Boy and combined motor cultivator and light tractor, was introduced by the sole concessionaire GW Wilkin at Kingston upon Thames in 1950. It had a 1.2 hp air-cooled JAP 2S four-stroke engine, a top speed of 3 mph and a dog clutch to engage the totally enclosed roller chain drive to its 13 in diameter Dunlop tractor tread tyres. Complete with a 10 in toolbar, two hoe blades, three grubbing tines, power take off pulley and engine toolkit, the original Farmers' Boy cost £52 ex works.

Optional accessories included steel wheels, a complete rotary scythe assembly for £8 10s 0d, a 12 in circular saw bench and frame for £13 10s 0d. An 8 in rotary cultivator was available at £7 14s 0d while a light twin roller with a seat was £9 16s 0d. The price list indicated that a shaft-driven hedge trimmer, a high-pressure pump and a spraying attachment were also in production. Later machines had a two- or four-stroke Villiers power unit.

At the 1951 Smithfield Show Raven Engineering introduced the Series II Farmers' Boy. This had an improved specification with the option of a 125 cc two-stroke or a 1 ½ hp four-stroke Villiers Mk 12 air-cooled engine. It had larger pneumatic-tyred wheels with stronger wheel hubs. The wheel track was adjustable between 10 and 22 in and the handlebars could be offset to either side of the tractor and adjusted for height.

1.41. GW Wilkin announced the arrival of the Farmers' Boy in 1950.

1.42. The Farmers' Boy light tractor.

Other improvements included a wider toolbar with provision for depth control and a power take-off belt pulley. In 1953 when the two-stroke engine was discontinued, new attachments included an 18 in cut lawn mower with a grass box, a 3ft cutter bar mower, a sprayer and a trailer.

The Series II Farmers' Boy cost £70 in 1956. The 10 in toolbar with side hoes and cultivator tines was an extra £3, a general-purpose plough was £23 and a 10 in rotary cultivator with eight blades was priced at £12 15s 0d.

The Major, the first of a new series of Farmers' Boy tractors introduced by Raven Engineering in 1957 was followed by the Minor, the Minorette and the Majorette. There were two models of the Farmers' Boy Major, the main difference being in the engine department. The Major Mk 15 had a 148 cc four-stroke Villiers Mk 15 engine and a 256 cc Mk 25 Villiers was used on the Major Mk 25 tractor. The engines were interchangeable and could be hinged backwards to allow the twin-pulley vee-belt drive to be changed from low to high speed or vice-versa.

Otherwise the Mk 15 and Mk 25 had the same specifications with a dog clutch to engage the enclosed roller chain drive to the wheels. Free wheel mechanisms in both hubs gave the user the choice of using the tractor with a fixed drive in forward or in the optional reverse gear, with drive to only one wheel freewheeling in either direction. The detachable 12 or 17 in cultivating rotor, chain driven from a power take-off on the side of the tractor, could be replaced with toolbar equipment, a front-mounted rotary scythe and cutter bar mower, a saw bench and other attachments.

The Farmers' Boy Minor rotary cultivator with a Villiers Mk 12 HS two stroke engine cost £48 when it was introduced in 1958. An alternative 129 cc four stroke Villiers engine was offered in 1960 and both engines were available for at least three years. There was a choice of a 129 cc Briggs & Stratton or a 120 cc Villiers engine for the self-propelled Minorette rotary cultivator when it was launched in 1959 but the Villiers option was discontinued in the early 1960s.

1.43. The Series II Farmers' Boy was launched at the 1951 Royal Smithfield Show .

1.44. There was a choice of a Briggs & Stratton or Villiers engine for the Farmers' Boy Majorette.

34

The Minorette had cleated solid rubber tyres, handlebars adjustable both sideways and for height and the two-speed belt drive gave speeds of ½ to 1 mph in low and 1 to 2½ mph in high gear. The 12 in wide rotary cultivator could be attached at the front or rear and additional blades were available to extend its working width to 16 or 27 in. Attachments for the Minor and Minorette included a toolbar with tines and hoe blades, a rotary mower, saw bench, hedge trimmer and a trailer.

The Farmers' Boy Majorette with the option of either a 4½ hp Briggs & Stratton or a Villiers Mk 12 HS four stroke engine and a three forward and one reverse gearbox appeared in 1962. An enclosed roller chain was used to drive the Majorette's 20 in diameter wheels which ran on fixed/free-wheel hubs within a speed range of ¾ to 6 mph and power take-off speeds of 1,500 to 3,000 rpm. Equipment for the Majorette, which was still being made in 1974, included a 20, 28 or 36 in rotary cultivator, a totally enclosed gear-driven 24 in rotary mower and a reversible plough.

Few changes had been made to the specification of Farmers' Boy Majorette and Minorette tractors manufactured by Bitex Ltd at Hampton Wick near Kingston-upon-Thames in 1964. Advon Engineers Ltd, also at Hampton Wick, were building the Farmers' Boy Majorette and the Minorette with a 129 cc Briggs & Stratton four stroke petrol engine in 1968.

The Farmers' Boy Little Giant rotary cultivator with a 4 hp Briggs & Stratton power unit, main drive clutch and two forward gears with maximum speeds of 1½ mph in low gear and 3 mph in high was added in the mid-1970s. Attachments included an 18 in rotary mower, a rotary cultivator, a plough, a rear tool frame with tines and hoe blades and a bulldozer blade.

Advon Engineers made an improved Little Giant and the Minorette in 1981. The Little Giant had a 5 hp four-stroke engine, three forward gears and one reverse and an adjustable wheel track. There was a choice of a 3 or 4 hp four-stroke petrol engine for the Farmers' Boy Minorette.

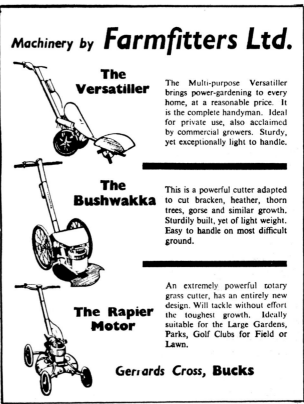

1.45. Farmfitters made the Versatiller in the 1950s.

FARMFITTERS

Farmfitters of Gerrards Cross in Buckinghamshire, which introduced the Versatiller powered gardening tool at the 1956 Smithfield Show, was equally well known for its power take-off mounted Jumbo air pump for inflating tractor tyres. The Versatiller was similar to the Gardenmaster power gardener with the tiller rotor in front of its 1½ hp Vincent two-stroke engine.

The Mk III Versatiller introduced in 1958 was advertised as a multi purpose rotary cultivator with various attachments designed to bring power gardening to every home at the reasonable price of £49 7s 0d. The Mk III Versatiller, described as the complete handyman ideal for private and commercial use, had a kickstart which sales literature suggested would be welcomed by ladies and elderly men who found other methods of starting an engine rather difficult.

Farmfitters made the Multigardner in the early 1960s. It was supplied with a complete set of gardening attachments and there was a choice of a 2½ hp two-stroke Clinton engine priced at £24 10s 0d or a 3 hp four-stroke Briggs & Stratton which cost £35 10s 0d.

The engine and tool frame assembly were mounted on the handlebars while various accessories could be attached to the tool frame in a few seconds. They included the Farmfitters Universal rotary cultivator, a heavy-duty Continental rotary cultivator, a lawn mower, a lawn edger and a sprayer. The choice of engines for the mid-1960s Multigardner included the same 2½ hp two-stroke Clinton engine used for the earlier model and a four-stroke 150 cc Villiers or a 2½ hp Clinton engine with a recoil or a press-button impulse starter.

FERRARI

The first Ferrari garden tractors and cutter bar mowers were made in the mid-1950s and within a few years this Italian company was also making four-wheel compact tractors. The single-cylinder Ferrari 74 rotary cultivator was imported by Norfolk-based JHB Implements in the mid-1970s. It had an 18 hp diesel engine, a single-plate dry clutch, six forward and three reverse gears, ground and engine speed power take-off, diff-lock and drum brakes. The ground speed power shaft was designed for use with a driven-axle trailer.

Ferrari garden cultivators available in the UK in the late 1970s included the Mk 32, the Mk 71 and the Mk 74. There were three versions of the Mk 32, with a

5 or 6 hp two-stroke petrol engine, a one, two or three forward speed transmission and a 70 cm rotary cultivator. The 10 hp petrol or diesel-engined Mk 71 had handlebars which could be rotated through 180 degrees. The Mk 74 with an 18 or 21 hp diesel engine, six forward and three reverse gears and a cultivating rotor with a minimum 56 cm and

1.46. Power from the Ferrari No 30 tiller's Briggs & Stratton engine was transmitted by a vee-belt and worm drive to the cultivating rotor.

1.47. Attachments for the 14 hp Ferrari 72S rotary cultivator included a plough, rotary and cutter bar mowers, spading machine and a potato lifter.

1.48. The Ferrari 90 motor cultivator handlebars could be rotated through 180 degrees.

respectively. The 1995 Ferrari catalogue included the 72S along with the 320, 330 and 340 cultivators with 6 to 10 hp engines and 20 to 32 in wide tiller rotors.

FLYMO

The Flymo GM and DM engine-over-rotor garden cultivators were introduced in 1978. The GM had a 2 hp Ducati two stroke engine with enclosed gearing to a 13¾ in wide tiller rotor. Advertisements pointed out that with its foldaway handlebars the GM would fit in a car boot without difficulty. The standard tiller rotor could, with additional blades, be extended to a maximum working width of 23 in.

The Flymo DM cultivator had a 3 hp Briggs & Stratton four-stroke engine to drive the tiller rotor which, depending on the number of tines used, had working widths of 12 to 19 in. The digging depth of 5 to 10 in was set with a four-position skid. With a pair of driving wheels on the rotor shaft the DM could be converted to a self-propelled tipping barrow.

Flymo widened its range of Flymo cultivators in 1980 when it bought the Westwood

maximum 97 cm working width, was sold either as a pedestrian-controlled rotary cultivator or a ride-on articulated tractor.

When Ferrari UK was established at Ely, Cambridgeshire in 1981 there were seven models of Ferrari garden cultivator, from the smallest Ferrari 30 with a 3 hp Briggs & Stratton engine to the 14 hp Ferrari 72S with an 80 cm tiller rotor. There was a choice of a petrol or diesel engine for the 72S which had a single-plate clutch, four forward and two reverse gears, independent brakes and an engine/ground speed power take-off.

Ferrari UK had moved to Oldham by the mid-1980s when there were eight models of Ferrari motor cultivator, ranging from the No 33 Motor Hoe with vee-belt drive to the cultivating rotor, to the 14 hp diesel-engined 72S with electric starter. Other models included the Ferrari 300, 80 and 90 series motor cultivators with 8, 10 and 12 hp petrol or diesel engines

1.49. Flymo GTM and GLM Tillermates had 5 hp Briggs & Stratton engines.

garden cultivator designs which, with a few modifications, were made by Westwood Engineering for Flymo. Two 3 hp and three 5 hp orange and brown Tillermates were included in the 1981 Flymo catalogue with prices starting at a modest £156 for the DM rising to £313 for the top of the range Flymo GTM.

The 3 hp DM and LM had 10 in wide cultivating rotors and 12 in rotors were used for the 5 hp DLM, GLM and GTM. A single forward speed was standard on all models with the exception of the Flymo GLM which also had a reverse gear. The professional GTM had five forward gears and one in reverse, the tiller rotor speed was variable from 70–150 rpm and with drive wheels on the rotor shaft it could be used to plough, ridge and harrow.

Early 1980s sales literature advertised Flymo-Norlett Commercial Products including the Flymo GLM and GTM Tillermates and the Flymo PowaSpade. The GLM and GTM with 10½ to 48 in digging widths had 5 hp Briggs & Stratton petrol engines and there was a choice of a 3 hp Briggs & Stratton or a 1,000 watt mains-electric motor with 38 m of cable for the PowaSpade.

The PowaSpade's 9 in wide front rotor, extendable with additional blades to 16 in, could be used to dig forwards, backwards or sideways. Rotating the rotor gearbox changed the direction of digging and with its various attachments the PowaSpade could also be used to hoe, weed, ridge and cut grass. Flymo cultivators were discontinued in 1984.

GARNER

Henry Garner, who established a car dealership at Nantwich, Cheshire in 1907, was importing farm tractors made by William Galloway at Waterloo, Iowa in 1918. The 29 hp tractor

1.50. The Flymo PowaSpade could dig forwards, backwards and sideways.

was sold under his name but the Garner was expensive and after poor sales it was discontinued in 1924. The company moved to Birmingham in 1926 to build lorries but the business failed and the new owners moved to Acton in London to manufacture lorries and other commercial vehicles. After World War II the company traded as Garner Mobile Equipment Ltd at Acton. The two-wheel Garner Light Tractor was advertised in January 1947 but the first tractors were not delivered until February 1948.

1.51. The first Garner two-wheel tractors were delivered to their new owners in February 1948. (Charlie Moore)

The pedestrian-controlled Garner had a 5–6 hp JAP petrol engine with an optional tvo conversion kit, a twist grip throttle on the handlebars, centrifugal clutch and a three forward and one reverse gearbox with a speed range of ½ to 8 mph. Power was transmitted through a differential and individual roller chain and sprocket drives from the half shafts to the wheels. Independent expanding shoe brakes on the half-shafts provided power steering. The standard Garner tractor on steel wheels cost £139 15s 0d in 1948, with optional pneumatic-tyred wheels adding £7 to the price.

The Garner de luxe, priced at £157, had an engine-mounted belt pulley driven by the power take-off, axle extensions and spade lug wheels with wide rims which could be split when working in narrowly spaced rowcrops. Garner implements included a floating toolbar made in collaboration with Stanhay, an off set Wilmot plough which enabled the user to walk on undisturbed ground instead of in the furrow, a trailer and bogie seat. A four-wheel tractor developed from the two-wheel Garner Light Tractor was introduced in 1949.

GRAVELY

The Gravely Model D with single wheel and wooden handles made by the Gravely Motor Plow and Cultivator Co. at Dunbar in West Virginia was introduced to American growers in the late 1920s. The first Model D tractors had a 2½ hp Indian motor cycle engine but most were sold with a 2½ hp Gravely four stroke air-cooled engine with pump-assisted splash-feed lubrication and started with a rope wrapped round the crank pulley.

The Model D did not have a gearbox but a 50:1 speed reduction was built into the gear drive transmission to the cleated steel wheel. The forward speed of 1 to 3 mph was controlled by a throttle lever on the handlebars. The American Gravely Model D specification stated that for export purposes the magneto, carburettor and air cleaner were accessories that would be fitted in the country of sale. Model D tractors sold in the UK were shipped in kit form to Gravely Overseas at Buckfastleigh in Devon where they were assembled and fitted with a Wico magneto and Zenith carburettor.

1.52. The Gravely Model D toolbar could be used in front or behind the tractor. (Roger Smith)

The Model D on the Gravely Overseas stand at the 1950 Royal Smithfield Show cost £80 complete with a toolbar which could be used in front of or behind the tractor. A toolkit with a ring spanner, oil can, screwdriver, adjustable spanner and starting rope was included in the price. A power-driven Gravely rotary plough, originally patented by Mr B F Gravely in

1.53. A front-mounted rotary plough was one of the many attachments made for the Gravely Model L.

America in 1936, had a cutter bar mower, a front-mounted mouldboard plough and toolbar equipment including cultivator tines, furrower and hoe blades were made for the Model D.

The Gravely Model L and Gravely X-Cel Estate Power Unit were also sold in Great Britain in the 1950s. The two-wheel Model L garden tractor with a 5 hp single-cylinder air-cooled Gravely engine, two forward and two reverse gears, differential steering and optional belt pulley, was designed for estate owners and market gardeners. Various front-mounted attachments for the Gravely L included a rotary plough, cultivator, roller, lawn mower, snowplough, circular saw, sprayer and a rotary brush. The four bladed rotary plough protected by a slip clutch and turning at 800 rpm ploughed a furrow 7 in deep and 12 in wide in light soil. Sales literature explained that the furrow would necessarily be somewhat narrower in hard ground.

The X-Cel Estate Power Unit with a single-cylinder 2½ hp air-cooled engine with dry sump lubrication, one forward and one reverse gear and a top speed of 3 mph was introduced in 1950. X-Cel attachments

1.54. The Gravely 8 hp 5260 Convertible could be used as a pedestrian-controlled or ride-on garden tractor.

included a plough, rotary cultivator, tool frame, sprayer, generator and a hedge cutter but it was equally suited for cutting grass with a 42 in front-mounted cutter bar or a cylinder mower.

Early 1960s Gravely tractors included the Junior Tiller, the Landworker and the Model L. There were also L1 and LS models with lower top speeds than the Gravely L. The all-British Gravely Junior Tiller with a 150 cc four-stroke Villiers engine, forward and reverse drive and a 27 in wide rotor under the engine cost £79 10s 0d when introduced in 1963. Engine power was transmitted by twin vee-belts to an idler pulley and then by flat belt to the tiller rotor. With alternative cultivating widths of 15 and 40 in, Gravely Tractors Ltd, based at Torquay and at Victoria Street in London, claimed the Junior Tiller would cultivate to a depth of 14 in when working in previously broken ground.

Accessories included a toolbar with tines, hoe blades and furrowers as well as a pair of pneumatic-tyred wheels for the rotor shaft when using the machine for rowcrop work. The Landworker, with two forward and two reverse gears, was built in the style of earlier Gravely tractors. Attachments included a 30 in front rotary mower, a 42 in front cutter bar, mouldboard and rotary ploughs, 20 gallon sprayer, circular saw and a power barrow.

Hurst-Gravely marketed the Gravely Convertible two-wheel tractor in the UK in the early 1970s. The two-wheeler was easily converted to a ride-on tractor with a sulky riding seat attached with a single pin and steered with the handlebars, while an alternative sulky riding seat had a steering wheel linked to the sulky wheels. An 8 or 10 hp Kohler petrol engine was used for the smaller models which had either two or four forward and reverse gears and a power take-off shaft.

The 12 hp Gravely Convertible tractor with a Kohler single-cylinder petrol or twin-cylinder diesel engine had four forward and a reverse gear and a two-speed power take-off. The range of front- and rear-mounted equipment included a front rotary, cutter bar or flail mower, a rotary plough, a rotary cultivator and a toolbar

Autoturfcare Ltd was the concessionaire for Gravely tractors in the UK from the late 1970s until 1982 when, following the acquisition of the Gravely

Motor Plow and Cultivator Co by the American Ariens Corporation, the tractors were marketed by Ariens UK Ltd from St Neots in Cambridgeshire. The mid-1980s range of Gravely two-wheelers included 8, 10 and 12 hp Kohler-powered Convertibles, the Pro Line series with Briggs & Stratton or Honda engines and the 16–19 hp Kohler-engined Pro Master series.

HOMELITE

Trojan Ltd, based at Croydon, which sold the tricycle-wheeled Trojan Monotractor and Toractor in the 1960s, marketed Homelite Roto-Tillers in the mid-1970s. The R8 Roto-Tiller, with an 8 hp Briggs & Stratton petrol engine, one forward and one reverse gear and a 23½ in wide tiller rotor, cost £169.95. The R5 with a 5 hp engine and 23 in rotor was £129.95 and the smallest 3 hp model with a 21½ in wide rotor was priced at £95.95.

HONDA

The world-famous motor cycle manufacturer introduced the Honda F60 and F190 pedestrian-controlled garden tractors to British growers in 1964. The 4 hp F60 had a single-cylinder Honda petrol engine while the twin-cylinder four-stroke air-cooled engined Honda F190 was rated at 7 hp when running at its maximum speed of 5,500 rpm. Both engines were started with a spring-loaded lever on the handlebars, which was pushed forward to load the mechanism before a sharp backward pull brought the overhead-valve power unit into life. Both tractors had an automatic centrifugal clutch, an oil-immersed multi-plate manual clutch for emergency stops and individual steering clutches for tight headland turns.

1.55. The Honda F-60 had a 4 hp single-cylinder Honda petrol engine.

1.56. A rotary cultivator, a crop sprayer and a water pump were among the attachments driven by the Honda F-190 power take-off shaft.

Engine power was transmitted by roller chain drive to a four forward and one reverse gearbox on the F60 while the F190 had six forward and two reverse gears. Both models had a top speed of 10 mph. The power take-off shaft on the F60 ran at 1,000 rpm and the cultivating rotor was replaced with a pair of pneumatic-tyred wheels when the tractor was used

1.57. The 8 hp Honda F 80 was made in the early 1970s.

with a reversible plough, rear toolbar equipment, a trailer and a water pump.

The F190 power take-off was used to drive a rotary cultivator, crop sprayer, water pump and other equipment. The wheel track was adjustable for rowcrop work and ploughing with a right-handed or reversible plough. The handlebars could be moved sideways to keep the operator's feet off freshly turned soil. The F60 cost slightly under £200 in 1964 while the basic F190 with a toolbar was £245 or £544 19s 0d with a rotary cultivator. Mudguards and headlamps were optional extras.

The F25 and F90 were added to the Honda garden tractor range in 1967. The F25, with a 2½ hp air-cooled four-stroke petrol engine and a 26 cm wide digging rotor, was designed for the smaller garden. The twin-cylinder, overhead-valve diesel

POWER TILLER
MODEL F600

A powerful lightweight designed to easily handle the full range of farming and gardening chores.

Work goes much smoother with the versatile F600. This rugged work-horse combines the handling ease and high reliability needed to make your agricultural activites easier. Plowing, weeding rotor tilling, harrowing, cultivating and other jobs are greatly simplified——even on hard-packed or sloping terrain. The F600 accepts a wide range of attachments for dry and wet field operation. A 5-horsepower, 4-stroke engine assures long-life dependability. It is linked to improved side clutches and a 3-forward/1-reverse speed transmission to provide a wider travel speed range. PTO enables you to power other equipment too.

HONDA

1.58. A six forward and two reverse gearbox and two-speed power take off were features of the Honda F600 introduced in 1980.

engine with electric or manual starting on the F90 was rated at 9 hp. The F90 specification included eight forward and two reverse gears, a four-speed power take-off and a rotary cultivator.

The 1973 Honda (UK) catalogue included the single-cylinder 2.8 hp F28, the 5 hp FS50 and the twin-cylinder 8 hp F80 Power Tillers with prices ranging from £140 for the F28 to £539 for the Honda F80.

A full range of attachments was made for the two forward speed F28 and the FS50 with a ten forward and one reverse gearbox and steering clutches on both wheels. The Honda F80 had a centrifugal clutch, six forward and two reverse gears, a power take-off shaft and a quick-detach rear-mounted rotary cultivator.

Four new Honda Power Tillers were introduced to the UK market in 1980. The 3½ hp F300 had two forward speeds provided by a high-low ratio vee-belt pulley and the chain-driven 50 cm wide rotor could be extended with extra tines to a maximum 75 cm working width.

The F400, also with a 3½ hp air-cooled four-stroke engine, had a vee-belt high-low ratio pulley and a two forward and one reverse speed gearbox. The F600, rated at 5 hp, and the 7 hp F800 had a vee-belt drive to the six forward and two reverse gearbox and both models had a two-speed power take-off.

The handlebars on the three larger models could be swung through 180 degrees allowing the operator to walk in front or behind the machine which was used with a variety of attachments, including a two-speed digging rotor, a dozer blade, a yard scraper, a toolbar and a trailer.

There were six Honda Power Tillers in the mid-1980s. The smallest 1 hp single-speed F110, suitable for the small garden, had a centrifugal clutch and vee-belt drive to a 20 cm cultivating rotor. Drive to the 2½ hp F210, 4 hp F315 and the 5½ hp F502, F510 and F660 cultivator rotor was engaged with a deadman's handle. A pair of transport wheels was provided

for the rotor shaft when the tractor was used for toolbar and trailer work.

The single-speed F210 was the smallest of the eight Honda Power Tillers available in the UK in the early 1990s. The 8 hp F810 with six forward and two reverse gears and a two-speed power take-off was the largest. Designed for commercial growers, the F810, like most Honda machines, could be used with slasher tines on a drum rotor up to 96 cm wide on the wheel axles, a 70 cm rear-mounted rotary cultivator and a full range of toolbar equipment.

HOWARD

Arthur Clifford Howard, an Australian engineer, established a company called Austral Auto Cultivators in 1921 to manufacture his own design of self propelled rotary hoe. Within a year he had made a self-propelled machine with a 60 hp engine and five rotary cultivating units with a total working width of 15 ft. Then, after adapting his rotary hoe for use with the Fordson tractor, Mr Howard turned his attention to pedestrian-controlled cultivators.

The first 6 hp petrol-engined Howard Junior Rotary Hoes were made in 1924. The Howard DH22, a four-cylinder 22 hp tractor designed for use with the Howard rotary hoe, appeared in 1928 and leading British implement manufacturers J & F Howard of Bedford exhibited a DH22 tractor at the 1930 Royal Show. AC Howard was entirely

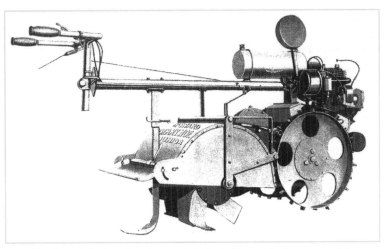

1.59. The Howard Junior rotary hoe made in Australia was the forerunner of the Howard Gem.

independent of the Bedford firm so when J & F Howard went out of business in the early 1930s the DH22 was still made and sold in Australia.

Howard made his way to Great Britain and in 1938 teamed up with Captain E N Griffiths to form Rotary Hoes Ltd in East Horndon, Essex. Captain Griffiths' connection with A C Howard had started ten years earlier when he saw the Howard rotary hoe and decided to import one for trial purposes. The trial was successful so Captain Griffiths arranged for J & F Howard to make about 100 rotary hoes, mainly for Austin, Fordson and Case tractors.

Rotary Hoes introduced the Rotehoe Gem in 1940 to replace the earlier Australian-designed machine. The pedestrian-controlled Rotehoe Gem Six, Eight and Twelve rotary cultivators were built in limited numbers at East Horndon throughout the war years when some components were hard to come by. In order to keep to the permitted level of production various makes of motor cycle engine, often obtained from scrap dealers, were renovated and used for Rotehoe machines.

Rotary Hoes Ltd acquired steam engine manufacturer John Fowler of Leeds in 1946 but within a year, and after introducing the FD22 crawler tractor, the company had sold the John Fowler business to Marshalls of Gainsborough in Lincolnshire. Production of the FD22 crawler, with a special drive arrangement for the Howard rotary hoe, was short lived but it was the forerunner of the Howard Platypus crawler made at Basildon between 1953 and 1956.

The 1940 Rotehoe Gem did not have a differential or a reverse gear but these refinements, together with swinging handlebars, were incorporated in the Howard Gem Series II introduced in 1947. Buyers could choose a Howard BJ (British Junior) or a JAP 6 hp petrol engine for early models of the Series II Gem but all later machines had a JAP power unit. Both engines had a dry sump and the moving parts were pressure lubricated with oil pumped from a reservoir mounted alongside the petrol tank.

The Series II Gem had a single-plate dry clutch controlled by a lever on the handlebars and a three forward and one reverse gearbox. A 20 in wide cultivating rotor was standard but 18 and 24 in rotors

1.60. The Series II Howard Gem could have a Howard BJ or a JAP four-stroke engine.

1.61. An air-cooled twin-cylinder engine was used for the Series IV Howard Gem.

1.62. Petrol and diesel engines were available for the Howard Super Gem Rotavator.

were also available. Production peaked at 200 Gems per week in the early 1950s. The firm was originally established at East Horndon but a decision by the local authority to alter the name to West Horndon in the early 1950s meant that the company changed its address but remained in the same premises!

The Series III and IV Gem Rotavators superseded the previous model in 1952. The Series III had a JAP 6 hp four stroke petrol engine to drive the land wheels and an 18 in or optional 20 in wide rotor. The more powerful Series IV with the choice of a 20, 24 or 30 in wide rotor had a 9.8 hp Howard air-cooled four stroke twin-cylinder engine with an optional two conversion kit to drive the 18 in diameter rotor at 172 rpm. Sales literature explained that the Gem Series III and IV were just the job for the smallholder because they had car-type controls with the gear lever in a four star quadrant on the handlebars, which were adjustable for height and lateral movement.

A Sachs diesel engine was added to the list of options for the Gem Series IV in 1957. The single-cylinder two stroke water-cooled Sachs engine

developed 9 hp at 2,000 rpm and was said to work for eight hours on a gallon of fuel. The specification included a diff-lock which was automatically engaged when the rotor drive was in gear. Although the chain-driven rotor was a permanent fixture, other attachments including a furrower and a roller could be hitched to the rotor cover. The Gem's power take off pulley was used to drive a soil shredder.

More than 50,000 Gems had been sold when the Series V appeared in 1961. The new model had a 12 hp Howard 810 cc twin-cylinder four stroke petrol engine, a twin-plate dry clutch and a three forward and one reverse gearbox with a top speed of 2.8 mph. There was a choice of a 20, 24 or 30 in wide cultivating rotor with a maximum working depth of 9 in.

In the late 1960s buyers were able to specify a single-cylinder Kohler 301T petrol engine or a 9 hp Sachs 500 diesel for the Gem. Both engines were available for the 20 and 24 in models but the 30 in Gem was only made with the Sachs 500 engine. A 10 hp Hatz ES 780 diesel or a Kohler K301T petrol engine were used for the Series V Gem in the early 1970s. Further improvements were made in 1976 with the introduction of the Standard Gem and Super Gem.

The Standard Gem was little changed except for a single-plate clutch and a higher top gear but the Super Gem was more robust with its strengthened rotor shields and more powerful engine options. The extra power for the Super Gem came from a 15 hp Kohler four stroke petrol engine or an 11 hp Hatz four stroke diesel engine. Sales literature explained that the Super Gem would stand up to the most brutal treatment.

1.63. The Howard Bantam made its debut at the 1949 Royal Agricultural Show.

The Howard Gem was made in many forms during its fifty-year production run. Some had large-diameter steel wheels for working in paddy fields and power units used over the years included four-stroke Howard BJ, JAP and Kohler petrol engines, four-stroke Hatz and Ruggerini and two-stroke Sachs diesel engines.

The Series I Howard Bantam with a 1¾ hp two stroke Mk 25C Villiers engine cost £71 10s 0d when it made its debut at the 1949 Royal Show. The slightly later Series II Bantam had a 2⅛ hp two-stroke Howard AC engine. Both had a two-speed gearbox and twin vee-belt pulleys providing four forward speeds, a power take-off and a two-speed rotor with the choice of a 10 or 14 in working width.

Controls on the laterally adjustable handlebars included a lever to select high or low gear, another to engage drive to the rotor, a throttle lever and a handle to lift the engine platform over centre to tension the vee belt drive from the engine. The rotor was easily removed, allowing the use of several attachments including a toolbar, sprayer, seeder unit, hedge trimmer, cutter bar and a cylinder mower.

The Howard price list for June 1956 included the

Series II Bantam with a 10 in rotor at £99 or £49 10s 0d down plus twelve monthly payments of £4 10s 10d. The Series I Bantam with the original 1¾ hp two stroke Villiers Mk 25C, a 1.9 hp four-stroke Villiers Mk 15 or a 120 cc four stroke BSA engine was £62 10s 0d. Existing Bantam owners were able to buy a Villiers four-stroke engine kit to upgrade Bantams with a two-stroke engine. Howard Bantams exported to America had a Clinton, Kohler or Briggs & Stratton engine and some Bantams with these engines were sold in the UK.

The Yeoman and Bulldog were added to the Rotavator family in 1954. The Yeoman, which cost £135, was advertised as the most advanced medium-powered cultivator available to the grower. There was a choice of a four-stroke 4.2 hp Villiers Mk 40 or a 4½ hp BSA petrol engine for the Yeoman with a single-plate clutch and two forward and one reverse speeds doubled up by a high/low ratio vee belt drive.

The Yeoman's two-speed 15 in wide rotor was attached by the 'Snaplock' single-lever quick-release clamping system which was also used to secure various optional attachments including front and rear toolbars, a cutter bar, a sprayer and a seeder unit.

The Howard Bulldog, which cost £49 10s 0d when it was introduced in 1954, had a JAP Model 2A 1¼ hp four stroke engine with separate vee belt drives to the wheels and to the 10 in or optional 15 in rotor. It was advertised as a true lightweight digger with self-sharpening hoe blades and ample power.

The Bulldog name became the subject of a dispute with the manufacturer of the German Lanz Bulldog tractor which resulted in the name being changed to the Howard Bullfinch. A sales leaflet at the time suggested that not so long ago mechanised gardening was a luxury for the few but the Bullfinch had put freedom from the spade and fork within the reach of all.

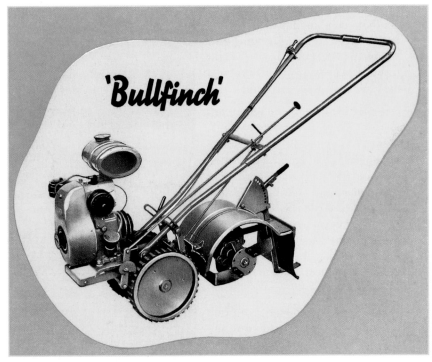

'Bullfinch'

1.64. The Howard Bullfinch was originally called the Howard Bulldog.

The Demon, with a 1.9 hp BSA four-stroke engine and a £77 10s 0d price tag, was launched at a Howard Rotavator dealer convention in 1957. A vee-belt transmitted power to the wheels and the 10 in wide cultivating rotor, which with additional blades, could be increased to a 15 in working width. The Bullfinch attachments were also suitable for the Demon and other optional equipment included side shields for the rotor, a power take-off and a belt pulley.

The combined total sales of rotary cultivators manufactured by Rotary Hoes and Clifford Aero & Auto Ltd during a ten-year period from the late 1940s exceeded 150,000 units. Clifford, which was making rotary cultivators at Birmingham in the late 1940s, changed its name to Clifford Cultivators Ltd and moved to West Horndon in 1956.

Within a couple of years Rotary Hoes had bought the Clifford assembly line and the rights to its garden cultivator range. Trailed and tractor-mounted Rotavators were sold under the Rotary Hoes name but the pedestrian-controlled range of Rotavators carried the Howard Clifford logo until the early 1960s.

In 1959 an improved range of garden Rotavators was launched at the Smithfield Show when the Howard Clifford price list included the Bullfinch at £73 7s 6d, the Bantam at £104 and the Gem at £240. The Demon, a more powerful version of the Bullfinch with a 2 hp Villiers engine, cost £81 10s 0d and two new models, the Howard Clifford 400 and 700, cost £175 and £215 respectively. The Howard-Clifford 400 was partly introduced to meet a demand from American rental yards and tool hire centres.

A 5½ hp Clinton petrol engine was used at first but later models had a 5½ hp JAP four stroke petrol engine, two forward gears and one reverse, adjustable handlebars and a 16 in wide rotary cultivator. An optional 6½ hp two-stroke German-built Hirth diesel engine became available in the early 1960s but by 1967 the 400 was powered by a 5¼ hp Kohler K161 petrol engine. The 400 could be supplied with an optional power take-off and pulley to drive a water pump, saw bench or hammer mill.

The more powerful Howard 700 had a 7 hp Villiers 28B two stroke petrol engine or a 7 hp two stroke Hirth diesel engine and a single-plate clutch. The two forward and one reverse gearbox combined with the new Selectaspeed unit with four gear ratios provided a total of eight forward speeds from 0.7 to 10.4 mph and four reverse speeds.

Other features included a diff-lock, two power take-off shafts with a belt pulley and fully adjustable handlebars that could be rotated through 360 degrees. There were four wheeltrack settings and the handlebars could be locked in a downward position on either side of the machine to serve as an axle stand to support the 700 when it was tipped sideways in order to remove a wheel.

The 15 in or optional 20 in wide cultivating rotor was secured to the power unit by the Howard single-lever 'Snaplock' quick-release implement attachment system which was also used to hitch a conventional or reversible plough, a toolbar and a cutter bar mower to the machine. H B Holttum & Co of Cambridge introduced a third wheel attachment for the 700 which could either be fitted to existing machines or purchased as the Holttum-Howard 700 three-wheel tractor. It was used mainly for haulage work.

The German Hako rotary cultivator with a 3 hp JLO engine was sold as the Howard-Clifford Hako in the late 1950s and early 1960s. There were six working widths from 6 to 42 in, obtained by adding or removing one or more slasher blades from each end of the cultivating rotor. Attachments for the Hako included a plough and a ridging body.

The Howard Rotavator production

1.65. The Howard-Clifford 400 had a Kohler petrol engine.

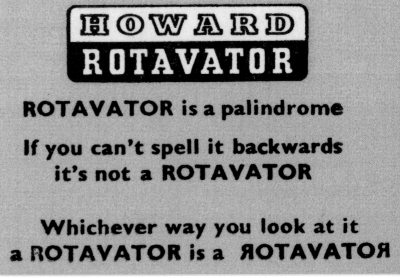

1.66. The word Rotavator is a palindrome, which reads the same forwards and backwards. It has become the standard term for rotary cultivators in the same way that JCB is the popular name for industrial diggers.

48

facilities were expanded in 1960 with additional premises at Harleston in Norfolk. This was followed in 1962 by a new forge and foundry at Halesworth in Suffolk. A new assembly plant was built in Ipswich and component manufacturing facilities were established at Washington, Tyne and Wear.

The closure of the West Horndon factory in 1974 coincided with Rotary Hoes changing the group name to Howard Machinery Ltd, which included Howard Rotavator Co, Howard Harvestore and Claas combine harvester importer J Mann & Son Ltd at Saxham near Bury St Edmunds, Suffolk. A new logo consisting of four Rotavator blades arranged to form a letter H was introduced and a new UK head office and spare parts depot was established at Saxham. This was closed in 1980 when the head office was relocated to Harleston .

1.67. The Howard 350 with a 5.3 hp Kohler engine replaced the Howard 300 in 1965.

The Howard Rotavator 300 introduced in 1961 with a 3 hp Kohler engine was somewhat underpowered so the Howard 350 with a Kohler 5.3 hp engine with an automatic decompression mechanism for easy starting was added to the Rotavator range in 1965. The models, apart from their different engines, were very similar with a cone clutch and a two-speed vee-belt drive to a two forward and one reverse gearbox. The land wheels were gear driven and a chain drive was used for the 16 or 23 in wide cultivating rotor.

The Howard 200, which made its debut in the early 1960s, was a modified version of the Bantam. The 200 had a 2

1.68. A new orange and green colour scheme was used for the Howard 352.

hp Villiers 1½ four stroke engine, two forward gears and, like the Bantam, it had a vee belt and worm drive to the rotor and the wheels.

A 1969 Howard Rotavator advertisement advised 'garden martyrs' to take a tip from the professionals and use a Rotavator to make light of all the muscle-aching and time-consuming jobs in the garden. It added that the Rotavator was a powered cultivator with powered wheels and when on the garden or going to and from the shed there would be no pulling, pushing, straining or swearing.

1.69. The Howard Dragon, launched in 1979, was one of the last models of Howard Rotavator.

The Howard 220 introduced in 1972 was a much improved version of the 200 with a 3.3 hp Aspera engine, clutch, diff lock and handlebars adjustable for height and lateral swing. A working width of 15 in was achieved by adding extra blades to the standard 10 in rotor.

The Howard Dragon, advertised as a two wheeled tractor power unit with a range of easy-to-attach implements, appeared in 1979. The 8 hp Kohler petrol engine transmitted drive through a two-speed vee belt to a gearbox with two forward speeds and reverse. There were two power take-off shafts, one to drive the 16, 24 or 36 in wide rotor and the other for a rotary or cutter bar mower attachment. Other Dragon implements included conventional and reversible ploughs, a furrower and a trailer.

The 352, which replaced the 350 in 1981, had a 5 hp Briggs and Stratton engine with a cast-iron cylinder and the angular engine cover above the familiar Rotavator orange chassis was painted olive green. A 1983 price list included the 352 at £800, the Dragon cost £1,500 and prices for the Gem ranged from £1750 to £2,538.

When Howard Machinery went into receivership in 1985, Dowdeswell Engineering of Stockton in Warwickshire bought the production rights for the Gem and 352 rotary cultivators and some of the Howard range of farm machines. Dowdeswell, which was not able to buy the Gem and Rotavator names, made the renamed Dowdeswell 352 and the Dowdeswell 650 – formerly the Gem – at the old Howard factory at Harleston until 1991.

ISEKI

The Iseki Agricultural Machinery Manufacturing Co was established in Japan in 1926 but the Iseki brand was not seen in the UK until Mitsui & Co of Denmark Street in London introduced Iseki four-wheel compact tractors in 1966 and the two-wheel KT600R Tiller in the following year. The green-painted pedestrian-controlled Iseki KT600R had a four stroke side-valve engine, ten forward and four reverse gears and adjustable handlebars.

Chain Saw Products Ltd of Manchester distributed five Iseki Tiller models between 1974 and 1979. They included the 3 hp two stroke KS280 garden tractor with two forward gears and one reverse and the 7½ hp KC450 two wheel tractor with four forward and four reverse gears and power take-off. A plough, cultivator, weeder, mower and a trailer were among the range of implements made for the Iseki tillers.

1.70. The Iseki KT600R tiller had ten forward and four reverse gears.

A Lely Iseki partnership was formed in 1979 to market Iseki compact and pedestrian-controlled garden tractors in the UK. The 1979 catalogue for Iseki Tillers, now in red livery, included the 11 hp KS650 rotary tiller with six forward and two reverse gears, the KS280, the KC450, the AC420 and the MC 1 mini rotary cultivator. The 7.3 hp KC450 tiller had four forward and four reverse gears and a power take-off to drive a rotary cultivator or rotary mower.

The AC420 tiller with a 5 hp two-stroke or optional four-stroke engine was supplied with two pairs of slasher rotors to give a 22 in tilling width. Designed for small gardens, the 2.2 hp MC-1 had a two stroke 2.2 hp engine above the 18 in rotor with a flexible air inlet tube clipped to the handlebars to reduce intake of dust. The 5 hp two stroke AC20 and 3½ hp four stroke AC40 rotary cultivators with an optional reverse gear and a 13 to 32 in wide cultivating rotor were added to the Iseki Tiller range in 1980.

The Lely Iseki partnership ended in 1986 when the Japanese manufacturer established Iseki UK at Little Paxton in Cambridgeshire. The new company offered British growers a

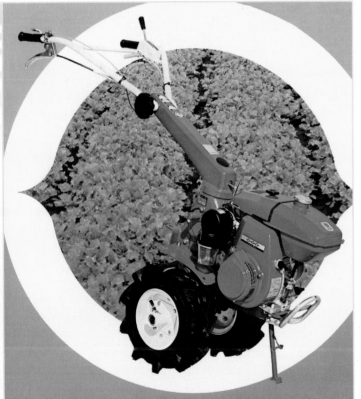

1.71. The 4 hp four-stroke Iseki KC450 power tiller was made in the late 1970s.

1.72. The Iseki KS280 tiller.

similar range of Iseki tillers and garden tractors with the usual choice of rotor widths, gear ratios and implements. The AC20 and AC40 Iseki rotary tillers and the 7½ hp A400 garden tractor were still listed in 1988 but they had been withdrawn from the British market by the early 1990s.

KINKADE

Although mainly used as a single-wheel garden tractor it was not a difficult task to convert the Kinkade to a two-wheel machine when added stability was required for rowcrop work. The Kinkade was driven by a system of gears with a four-stroke engine mounted in the centre of the wheel.

With an optional second driving wheel and axle extension the Kinkade was claimed to plough a furrow 8 in wide by 6 in deep. In two-wheel format it could be used astride the row with a 6 ft wide toolbar which had a castor wheel at each side of the bar for cultivating, hoeing, ridging and lifting potatoes. Other Kinkade implements included a disc harrow, a fertiliser spreader and seeder units.

1.73. The Kinkade garden tractor was made in the late 1940s.

KUBOTA

A company founded by a Mr Gonshiro Kubota in 1890 to make iron castings eventually became the Marubeni Corporation, a general engineering company with diverse interests including shipbuilding, marine engines and irrigation equipment. The first Kubota stationary engines were made in 1923, the first garden rotary cultivators appeared in 1948 and the first Kubota compact tractors were introduced in 1960.

John Croft of Whitley Bridge in North Yorkshire introduced six models of two-wheel Kubota garden tractor, ranging from the 2½ hp Power Tiller T120 to the 5½ hp T720 in 1978. The smallest single-speed Kubota T120 with an 18 in rotor cost £195, with rotor extensions available at extra cost. The more versatile 2½ hp T320, which also came with a cultivating rotor, had two forward and two reverse gears. Like the T120 it was an engine-over-rotor type of rotary cultivator.

The specification for the 3½ hp T420 and T420C included a three forward and two reverse gearbox, two-speed power take-off and a 26 in wide cultivating

1.74. The 2½ hp single-speed Kubota T150 tiller could, with the handles folded, be carried in a car boot.

rotor. Owners of the T420C enjoyed the added advantage of steering clutches in both wheel hubs. The four forward and two reverse gear T620 with a 4½ hp engine and the 5½ hp T720 with six forward speeds

and two in reverse completed the 1973 range of Kubota Power Tillers. The T620 and T720 had a 37 in wide cultivating rotor and, with the exception of the T120, all Kubota Power Tillers could, with optional traction wheels on the rotor shaft, be used for ploughing, cultivating, ridging and grass cutting.

The introduction of the new T250, T350 and T450 Power Tillers in 1982 coincided with Kubota's move to larger premises at Thame in Oxfordshire. The single-speed T250 and the two forward and one reverse speed T350 had 3½ hp Kubota engines. The 4½ hp T450 Power Tiller, designed for the professional grower, had the same gearbox as the earlier T720 and, complete with a rotor extension kit, cost £484. The lightweight single-speed T150 with a 2.2 hp four-stroke engine was added to the Kubota T series of power tillers in the mid-1980s. It had a 24 in tilling width and could be folded for transport in the boot of a family car.

The petrol-engined AT60, AT70 and AD70 diesel Kubota garden tractors for commercial growers were introduced in 1988. The AT60 had a 20 in rotor while the AT70 could be supplied with a 20 or 26 in rear-attached rotary cultivator with an unusual forward and reverse drive to the tiller rotor.

The AT60 had two forward gears and one reverse but the AT70 and AD70 were equipped with a three

1.75. The Kubota AT70.

1.76. The Kubota T350 power tiller.

forward and one reverse gearbox and steering clutches. Both models, together with the AT55 with a 5.2 hp engine, four forward and two reverse gears, were still available in the mid-1990s.

LANDMASTER

Byron Horticultural Engineering of Hucknall near Nottingham made the first two-wheel Landmaster garden tractors (page 25) in the late 1940s. The Gamecock, Kestrel and Hawk rotary cultivators with Villiers engines and Albion gearboxes were current when the Hucknall company changed its name to Landmaster Ltd in the early 1950s. The 2 hp Gamecock with three forward speeds and a 10 in wide cultivating rotor cost £75.

The Mk I Kestrel de luxe with a 4 hp engine and a three forward and one reverse gearbox was priced at £115. The 14 in cultivating rotor secured with four bolts was easily removed when the Kestrel was required for inter-row hoeing and other toolbar work.

The Mk II 3 hp Kestrel, which cost £100, also had three forward gears but its 14 in cultivating rotor was a permanent attachment. The Hawk, which cost £145, had a 5 to 6 hp Villiers engine, three forward gears and reverse and an 18 in cultivating rotor.

Following its acquisition of Dashwood Engineering in the mid-1950s Landmaster made the Landmaster Colwood Mk II Model B Landmaster L40 and the improved Mk I Hawk. The Landmaster Mk I Hawk had a 5 hp Villiers petrol or a 5½ hp BSA two engine, a quick-release 18 in wide cultivating rotor, three-speed power take-off and twin shock absorbers to protect the transmission. The Mk II Model B with a 1.3 hp Villiers Mk 12 four-stroke engine, two forward gears and enclosed roller chain drive to the single-wheel was suitable for rowcrop work, grass cutting and hedge trimming.

Landmaster was part of the Firth Cleveland Group when it introduced two models at the 1956 Royal Smithfield Show. The larger of these, the Landmaster Eagle, had a 7½ hp Bradford-Sanders diesel engine. Equipment for it included a rotary cultivator, an 18 in rotary mower and a hedge trimmer.

Its companion, the Gardenmaster, with its 1 hp two-stroke engine and a forward tool-head, was described as an all-purpose tool for the farm, nursery and private

1.77. The diesel-engined Kubota AD70 was introduced in 1988.

1.78. The 2 hp Landmaster Gamecock cost £75 in 1954.

garden which could dig, hoe, cut grass and trim hedges. Complete with a cultivating rotor, universal tool head with tines and hoe blades and a spin weeder, the Gardenmaster cost £49.

The Gardenmaster's optional flexible drive shaft could be used to power a small circular saw, drilling machine, lathe and an air compressor for paint spraying, etc. It could be used to dig in either direction by turning the rotor head through 180 degrees to reverse the direction of the digging blades and walking backwards with the machine avoided the need to tread on the freshly cultivated ground. Users were advised that the best results would be obtained by setting the digging rotor at the required depth and swinging it from side to side in pendulum fashion as the machine moved along.

Optional equipment included a single-wheel conversion unit for use with a ridging body and a 7 in rotor for working in narrowly spaced rows. Additional rotor blades were available to increase the standard 11 in digging rotor to a working width of 18 or 24 in.

The Landmaster L150 rotary cultivator cost £95 when it made its debut at the 1959 Chelsea Flower Show. Power from a 4 hp two-stroke JAP 158 cc engine, which used between 1 and 3 pints of petrol an hour, was transmitted by a two-speed vee-belt drive and an oil-immersed roller chain to the cultivating rotor.

Two, three or four sets of curved, square, scimitar or pick tine blades could be used on the rotor to give a 12 to 44 in digging or cultivating width. With a pair of pneumatic-tyred traction wheels on the rotor shaft the L150 was ready for use as a garden tractor with a reversible plough, toolbar equipment, a seeder unit, a 33 in wide rough grass cutter, a centrifugal water pump, a light trailer and other attachments. A rotary ridger, mainly for export, was introduced for the L150 in 1961. It had left- and right-handed curved blades for ridging up potatoes and for adding protective mulches to the base of growing crops.

The two-stroke Landmaster 150 with rubber-tyred land wheels cost £114 10s 0d in 1962, an optional 4 hp four-stroke JAP engine adding £9 to the price. A 148 cc JLO two-stroke engine was used in the late 1960s but sales literature explained that alternative engines were available.

The ¾ hp two-stroke Gardenmaster 34 and 1½ hp Gardenmaster 80 rotary cultivators were also

1.79. The Landmaster L150 with a two-stroke engine was introduced in 1959.

exhibited at the 1959 Chelsea Show. Complete with a front-mounted 11 in cultivating rotor and a spinweeder, the Gardenmaster 34, which cost £37 10s 0d, could with various optional attachments be used for all sorts of gardening tasks from digging and hoeing to trimming hedges and mowing lawns.

1.80. The Landmaster 243 engine could be positioned over the engine for deep digging or over the wheels for lighter duties.

There were four models of the royal blue Gardenmaster Power Gardener with white wheels in the early 1960s. The smallest Gardenmaster 34 cost £39 0s 0d and the Gardenmaster 80 with a JAP 1½ hp two stroke engine, which was popular with vegetable gardeners, was £49 15s 0d. Attachments included a flexible engine-driven cable for a small hedge trimmer or a drill chuck for the handyman to drill holes in wood and metal.

The 3 hp four stroke Gardenmaster 85 with a streamlined engine cover cost £52 10s 0d and the 3 hp de luxe Gardenmaster 95 priced at £62 10s 0d introduced in 1961 completed the range. A press report described the de luxe 95 as an uncommonly smart cultivator with the engine completely encased in a streamlined blue cowling - definitely the machine to keep its owner one step ahead of the Joneses. Standard equipment included a spin weeder and 12 in wide digging rotor with extra blades to increase its working width to 18 in.

Optional attachments for the Gardenmaster Power Gardener range included a grass mower, water pump, Allman sprayer and a flexible drive for Tarpen or Heli Strand Tools hedge trimmers.

The Gardenmaster 100, introduced in 1962, had a 3 hp four stroke Briggs & Stratton engine and a lever-operated clutch to disengage drive to the 12 or 18 in wide digging blades on the front rotor. It was the largest member of the Landmaster Power Gardener family.

An advertisement at the time recommended the use of the Gardenmaster as a remedy for 'clay diggers' cramp' which was said to be particularly prevalent among residents of new housing estates, the symptoms being a difficulty in getting feet off the ground and ground off the feet.

The 1968 Landmaster range comprised the Model 66, 88 and 100 front rotor cultivators and the engine-over-rotor Model L120, L150 and L243 machines. The Model 66 with a 2½ hp two-stroke Aspera engine and a pair of slasher units with a 12 in cultivating width was £56 10s 0d and the 3 hp Model 88 with a 3 hp four-stroke Aspera and an 18 in cultivating rotor cost £66 10s 0d. The Landmaster Model 100 with an 18 in wide slasher rotor and a hand-operated clutch to disengage the rotor drive when starting a 3 hp Briggs & Stratton engine cost £76 10s 0d.

1.81. Spanners were not required when changing tool heads on the Gardenmaster Power Gardener 34.

Accessories included a rotary grass cutter, a rotary lawn rake and a third wheel conversion set which was required when using the Landmaster 66, 88 or 100 with hoe blades, a ridging body or a spin weeder. Landmaster-approved equipment included the Tarpen power take-off driven flexible drive shaft for hedge cutting, sawing and pumping water.

The 3 hp Landmaster L120 had a four-stroke Briggs & Stratton engine, hand clutch control, a 24 in wide digging rotor, stabilising transport wheels and the usual range of implements and accessories including a rotary grass cutter, a plough and a toolbar. The Landmaster L150 rotary cultivator with a 158 cc JAP engine, two-speed vee-belt drive, reduction gears and roller chain drive to the 33 in wide cultivating rotor cost £123 0s 0d in 1964. Four years later the price of the L150, now with a 6 hp two-stroke JLO engine, had risen to £142 10s 0d.

The specification of the L243 included a 4 hp four stroke Briggs & Stratton engine, two forward gears and one reverse, a 12 in wide rotor with optional extensions giving a 36 in cultivating width, and adjustable handlebars. It was advertised as two machines in one 1986 catalogue because the engine could be moved forward over the cultivating rotor for added penetration or backwards over the wheels and used as a garden tractor for hoeing, ridging and other rowcrop work.

The 2½ hp two stroke Landmaster Mo'dig, similar in appearance to the Gardenmaster 80, was advertised in the mid-1960s as a new dimension in power gardening. As the name suggests, the Mo'dig could be used as a rotary mower or a power cultivator. Complete with mower, grass collecting bag, rotary lawn rake, spin weeder and front digging head it cost £69.

Landmaster was making the L66, L88 and L120 rotary cultivators and the 3 hp Lion Cub with one pair of slasher rotors when it moved to Poole in Dorset in 1971. There was a noticeable decline in the sales of pedestrian-controlled machines in the late 1970s when a reduced range of Landmaster machines included the L130, the L140, the Lion, the Lynx and the forward tool-head Lion Cub and L88 Super.

The engine-over-rotor L130, L140 and Lion had Briggs & Stratton 3 hp, 4 hp and 6 hp two-stroke or 7 hp four-stroke engines respectively and with driving wheels on the rotor shaft were used with the usual selection of toolbar accessories. A 19 in cut Stoic rotary mower unit was available for the L130 and L140 cultivators. The Landmaster Lynx was a 5 hp self-propelled rotary cultivator with forward and reverse gears and an 18 in wide rear tiller rotor. The 3 hp Briggs & Stratton-engined Lion Cub and L88 Super were the latest in the long line of Gardenmaster forward tool-head cultivators introduced in the mid-1950s.

The Landmaster division of Boscombe Engineering, also based at Poole, was marketing the L88 and L120 along with the Landmaster-Gilson ride-on tractors in 1975. The E10 mains-electric Gardenmaster digger with a 12 in wide digging rotor was made at Poole in the late 1970s. Wolseley Webb acquired the Landmaster business in 1980.

1.82. The Landmaster Mo'dig could be used as a rotary digger or a rotary mower.

MAYFIELD

Harry Whitlock, an engineer with Croft brothers of Croydon during the mid-1940s, designed a cheap pedestrian-controlled cutter bar mower. In 1949 Mayfield Engineering at Croydon introduced the Mayfield Mk 10 garden tractor based on the Whitlock design. Designated Croft Mayfield garden tractors, they were made by Mayfield Engineering until John Allen & Sons at Oxford acquired the Mayfield and Croft businesses in 1965. Production of some Mayfield Croft models continued under the Allen Mayfield name and an additional range of Allen Mayfield garden tractors was added in 1971. When trading as Allen Power Equipment in the early 1970s they sold the original Croft garden tractor designs to

Arun Tractors but the Mayfield name and Allen Mayfield garden tractor designs were not included in the deal. The new owners, trading as Riverside Precision Ltd. at Rustington near Littlehampton, made some of the existing Allen Mayfield models under the Arun name along with a new range of Arun garden tractors. Serious production ceased in 1981 but the last Arun tractor was not sold until 1991.

The Mk 10 Mayfield made between 1949 and 1952 had a single-plate clutch and a three forward speed Albion motor cycle gearbox with a top speed of 3½ mph. Later models could be supplied with an optional reverse gear. The 1.2 hp four-stroke Villiers engine was started with a hand lever attached to the standard motor cycle kickstart pedal or with a rope wrapped around the crank pulley. A roller chain transmitted power to the wheels with free-wheel mechanisms in the hubs. Twin vee-belts were used to drive the detachable 36 in wide central or offset mower cutter bar.

1.83. A mower cutter bar was one of the attachments for the Croft Mayfield Mk 12 garden tractor made from 1953 to 1962. (Bill Castellan and Ray Smith)

A rear toolbar with cultivator tines, hoe blades, ridging bodies, two Jalo seeder units and a mouldboard plough were made for the Mk 10 Mayfield. Other attachments included a front-mounted dump box, a saw bench, an air compressor for paint spraying and inflating tyres and a small 110 v DC generator for use with a hand-held electric hedge trimmer.

To meet the demand for more powerful garden tractors Mayfield Engineering of Dorking, Surrey introduced four new models with four stroke engines, three-speed gearboxes and optional reverse gear in the 1950s. The 1¾ hp Croft Mayfield Mk 12 appeared in 1953 followed by the Mk 15 in 1955, the Mayfield Wentworth in 1957 and the 3 hp Mk 20 in 1958. Later models of the Mk 15 and Mk 20 had 3 hp Villiers and 4 hp Briggs & Stratton engines respectively.

The Mayfield Wentworth was an adaptation of the Mayfield Mk 12 but with lower forward speeds and a second power take-off to drive a 10 in rotary hoe attachment. The standard 22 in wheel track could be extended to 26 or 30 in by reversing the land wheels but with its 10 in wide rotary hoe, driven by a heavy-duty motor cycle roller chain from the power take-off, the Wentworth always left tyre marks on the cultivated ground.

An advertisement in 1959 informed potential customers that ploughing conserved the soil and suggested that the Mayfield would also conserve their energy and save both time and money. Another advertisement pointed out that the Mayfield tractor was of robust construction yet light to handle and suitable for ladies to use. Fourteen attachments for Mayfield tractors included a plough, various cultivating implements, a potato lifter, a front cutter bar mower, a wheelbarrow and a snow plough.

The Mk 14 with a 2½ hp Briggs & Stratton engine replaced the Mk 12 in 1961 when the Mk 21, Mk 25 and the Hoe & Mow were added to the Mayfield range of garden tractors. The Mk 21 with a 3½ hp Briggs & Stratton engine and a 21 in rotary or a 24 in cylinder mower was the only Mayfield tractor made purely for cutting grass. The Mk 25 had a 4 hp Villiers engine, a three forward and one reverse Albion gearbox, a top speed of 3½ mph, adjustable handlebars and a 3 ft centrally mounted or offset cutter bar.

The single-wheel 2 hp Hoe & Mow designed for inter-row hoeing and cultivating could also be used with an 18 in front-mounted rotary mower and an 18 in saw bench. The two-speed gearbox had top speeds of 1½ mph and 2½ mph and two heavy-duty motor cycle chains transmitted drive to the 14 in diameter pneumatic-tyred wheel.

The Mk 16, introduced in 1962 and made for three years, was the last new model of Mayfield garden tractor. Apart from its 3½ hp Briggs & Stratton engine with a hand lever or recoil starter it was identical to the Mk 14.

The lower price of Mayfield tractors in the early 1960s was having an adverse effect on sales of Allen Scythes, so John Allen & Sons decided in 1965 to buy Mayfield Engineering and the manufacturing rights for Mayfield garden tractors. Production of the Mayfield Mks 15, 16, 20, 21, 25 and the Hoe & Mow continued under the Allen ownership and although the Mk 21 was discontinued in 1968, the others were made until 1971.

The Allen Mayfield Mk 7 and Mk 8 garden tractors, better known as the Allen Mayfield Seven and Allen Mayfield Eight, were introduced by John Allen & Sons in 1971, remaining in production until 1985. The Mayfield Seven had a 7 hp Briggs & Stratton engine while an 8 hp Kohler was used for the Mayfield Eight. Both had a recoil starter, three forward gears with a top speed of 4.1 mph and an optional reverse gear.

1.84. The Mayfield Hoe & Mow. (Bill Castellan and Ray Smith)

1.85. The Allen Mayfield Eight was introduced in 1971. (Bill Castellan and Ray Smith)

Attachments included a front-mounted 30 in cylinder mower, a 2, 3 or 4 ft wide cutter bar and a twin-blade 32 in rotary mower. Most rear-mounted Mayfield implements including the plough and toolbar, dozer blade and trailer were suitable for the Allen Mayfield Seven and Eight and a range of Sisis turf care equipment was also made for the tractors.

A hydrostatic version of the Allen Mayfield Eight with a Peerless reduction gear unit and an Eaton Model 6 hydrostatic transmission was added in 1980 and made for the next three years. The hydrostatic Mayfield Eight had an infinitely variable forward speed to a maximum 5 mph and 1 mph in reverse. The four-wheel ride-on Mayfield Merlin (page 252) was also made during the Allen Mayfield era.

1.86. The Arun tractor. *(Bill Castellan and Ray Smith)*

Allen Power Equipment sold the original Croft Mayfield designs to Riverside Precision and Sheet Metal Ltd at Ford near Arundel and later at Littlehampton in the early 1970s. Since the Mayfield name was not included in the deal the new owners adopted the Arun name. Allen Power Equipment continued production of the Allen Mayfield Seven and Eight for another ten years but the Hydrostatic Eight was discontinued in 1983.

Riverside Precision and Sheet Metal introduced the new Arun garden tractor with the choice of a 3 hp or 6 hp Villiers or a 5 hp or 7 hp Briggs and Stratton engine in 1975. Power was transmitted through clutched twin vee-belts, initially to an Albion

gearbox but this was eventually replaced by a new Arun three forward and one reverse speed gearbox.

Later Arun tractors were supplied with a 5 or 8 hp Briggs & Stratton or an 8 hp Kohler engine but towards the end of its production run customers could specify almost any make of engine for their tractor. Front-mounted cylinder, rotary and cutter bar mowers together with a range of rear-mounted attachments similar to those for the Allen Mayfield were made for the Arun.

The original single-wheel Croft Hoe & Mow was made by Riverside Precision & Sheet Metal until the stock of parts bought in the early 1970s was exhausted. A new single-speed Arun Hoe & Mow with a full range of accessories was introduced in 1978. It was very similar to the earlier model but it had a 4 hp Briggs & Stratton engine and vee-belt drive to the front-mounted cutter bar or cylinder mower.

The Arun Hoe & Mow had a speed reduction unit attached to the engine instead of the gearbox as in the past. An optional two-wheel conversion with the wheels attached to a shaft through the hollow centre of the single wheel axle gave the Hoe & Mow added stability when it was used with a cutter bar or cylinder mower.

Regular production of Arun tractors came to an end in 1981 but they were still made in small numbers and customers were able to specify almost any make of engine. The Hoe & Mow and the Arun tractor with an 8 hp Kohler or 5 hp BSA engine and a cutter bar mower, rotary mower or hay sweep were still available in 1986 and the very last Arun machine was made in 1991.

MONRO

Geo Monro of Waltham Cross in Hertfordshire entered the single-wheel Monro Monotrac, two-wheel Monro Duotrac and the Simar Rototiller 5 in the 1930 World Agricultural Tractor Trials held in Oxfordshire. The market garden cultivator trials, which involved inter-row cultivation work, were held separately on a market garden at Pitchill near Evesham in Worcestershire.

The Monotrac with a single-cylinder two-stroke engine air-cooled by vanes on the flywheel, a Villiers carburettor, a 1¼ gallon petrol tank and a throttle control lever on the handlebars, had a top speed of 3 mph. The 22 in diameter wheel with self-cleaning V-shaped strakes was supplied with

1.87. The Monro Monotrac was entered in the 1930 World Agricultural Tractor Trials.

a road band. The Monotrac, which cost £50 complete with cultivating equipment, developed 3 hp at the drawbar with the engine at its maximum speed and the 4 in diameter side-mounted belt pulley was rated at 3½ hp.

The single-speed Duotrac motor cultivator with two 24 in diameter and 3 in wide steel wheels and also self-cleaning V-shaped cleats cost £38 complete with cultivator tines and hoe blades. The single-cylinder two-stroke air-cooled engine had a Villiers carburettor and one-gallon petrol tank. When running at its rated speed of 1,500 rpm the engine developed 2¼ hp at the drawbar and 2¾ hp at the front-mounted 3¾ in diameter belt pulley.

The Swiss-built two-wheel Simar Rototiller 5 garden tractor introduced to the UK market in 1929 was also entered in the 1930 World Tractor Trials. The

Rototiller 5, which cost £98, had a 4½ bhp air-cooled two-stroke engine, two-speed gearbox with top speeds of ¾ and 2 mph and a 5½ in diameter rear-mounted belt pulley. Geo Monro imported the Swiss-built Simar garden tractors for the next twenty years.

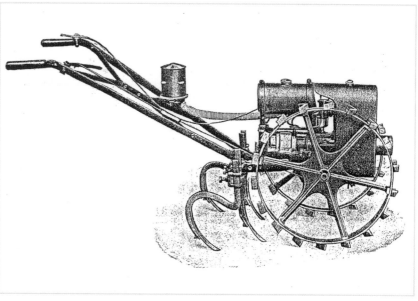

1.88. The Monro Duotrac with cultivating tines and hoe blades cost £38 in 1930.

1.90. The 8 hp British Simar 56 was the largest Rototiller in the Geo Monro catalogue for 1952. *(Roger Smith)*

1.89. Made by Geo Monro in 1953 the 5 hp British Simar in low gear could till an acre in ten hours. *(Roger Smith)*

The Waltham Cross company was granted a licence in 1948 to build the Simar Rototiller 56 to its Swiss specification. At this time it cost £180 on steel wheels, pneumatic-tyred wheels adding £15 15s 0d to the price. The British Simar 35 Rototiller, made by Geo Monro from 1950 to 1958, had a two-stroke 4 to 5 hp Simar engine, two forward gears and an 18 in cultivating rotor usually referred to as the miller on Simar Rototillers.

The engine air cleaner consisted of an oil bath pre-cleaner with wood shavings in the tube connecting it to the carburettor. Owners were warned that the engine would be damaged if the shavings became soaked with oil or were too dry. In low gear the Rototiller 35 cultivated an acre to a maximum depth of 12 in in ten hours and in high gear it could till an acre up to 6 in deep in half that time.

1.91. The hinged Simar badge concealed the two-stroke engine's sparking plug. *(Roger Smith)*

The 8 hp British Simar 56 and 56A were the largest Rototillers made by Geo Monro between 1948 and 1961. The two-stroke Simar engine, still with an air cleaner using oil-soaked wood shavings, needed about four gallons of petrol to plough an acre in eight hours. The 56 had two forward gear ratios, selected by changing the position of the driving peg in both wheel hubs.

The 56A had the added advantage of two reverse and two forward gears, the high/low ratio was selected with the wheel hub pegs and reverse was engaged with a lever on the handlebars which could be offset to either side of the machine. Another lever was used to engage the drive to the standard 20 in miller or optional 14 and 26 in wide millers.

The 1952 range of Monro two stroke engined garden tractors included the 5 hp Rowtrac (see page 70), the Rototiller 35 and the Rototiller 56, together with the Series II and III Monrotiller rotary cultivators which had been introduced in 1951. The Series II Monrotiller could have a 1¼ hp Villiers two stroke engine or a 1 hp four-stroke engine, which with pneumatic-tyred wheels with a 12 in miller and a toolkit, cost £80 ex works.

An improved Series II Monrotiller, introduced in 1952, had a 1.7 hp Villiers Mk 12 four-stroke engine and a two-speed gearbox. Monrotiller attachments included a 25 gallon sprayer with a 6 ft spray bar and hand lance, a 5 cwt tipper truck and a rear-mounted rotary scythe belt driven from the power take-off. The Series II with a front toolbar cost £64 or £80 when supplied with a 12 in wide spring-tined miller rotor or £86 with a front-mounted cutter bar.

A 2¼ hp Villiers Mk 15 four-stroke engine,

1.92. The Series II Monrotiller cost £80 in 1954. (Roger Smith)

1.93. Made in 1962, the Series III Monrotiller had a 2¼ hp Villiers engine. (Roger Smith)

the only overhead valve engine made by Villiers, was used for the Series III Monrotiller. Some Series III tractors had a two forward and one reverse gearbox while others were sold with three forward and three reverse gears. The handlebars could be offset to either side to avoid walking on freshly cultivated ground and four types of tine were available for the 14 in wide miller rotor. Attachments for the Series III Monrotiller, which cost £120, included cutter bar, rotary and cylinder mowers, a toolbar, a sprayer, a saw bench, a pump and a generator to power an electric hedge trimmer. The 2¼ hp Series IIC Monrotiller with the same Villiers engine had a two-speed gearbox and ratchets in the wheel hubs could be set for a positive drive or freewheeling.

The Monrotiller Minor with a 98 cc two stroke engine cost £59 when it was introduced in 1959. It was an engine-over-rotor type garden tiller with 15 or 32 in wide tiller rotors and looked rather like an outboard motor on wheels with a pair of handlebars and cultivating rotor. Other attachments for the multi-purpose Monrotiller Minor included a rotary mower, a ridging body, hoes, a sprayer, a small trailer and a Tarpen hedge trimmer. Unusually for a small machine, it had a clutch to engage drive to the tiller rotor and the handlebars could be offset to either side of the machine.

A more powerful Monrotiller Minor with a Mk. IA Villiers 7F 2¼ hp two-stroke engine appeared in 1962 when the Monro Transport Co Ltd, a subsidiary of Geo Monro, advertised the mighty Monrotiller with a 14 in rotary cultivator. The Series IIC was £108 ex works and the Series III cost £120, though hire purchase terms were available.

A garden machinery directory for 1973 listed the Series III Monrotiller, the Crunchtiller and three models of the Rototiller sold by Mechanised Gardening Ltd of Great Gransden in Bedfordshire. The Rototiller 8 had an 8½ hp petrol or 8 hp diesel engine, six forward and two reverse gears and a 20, 24 or 28 in cultivating rotor. A 10 hp two-stroke petrol or 8 hp two-stroke diesel engine provided the power for the Rototiller 10 which had a four forward and four reverse gearbox and a 24, 32 or 38 in wide rotor. A two-speed rotor for deep digging with a 32, 38 or 42 in wide rotor was a special feature of the 12 hp Rototiller 12 with either a two-stroke or four-

stroke diesel engine and a four forward and reverse gearbox.

The early 1970s wheel-driven Series III Monrotiller specification included a 3½ or 4½ hp four-stroke engine, three forward and three reverse gears and a 14 in rotary cultivator with four types of easy-fit tines. Other attachments included a rotary scythe, a cutter bar and a trailer. The engine-over-rotor Crunchtiller rotary cultivator also came with a 3½ or 4½ hp four-stroke power unit and a 15 in rotor extendable to 28 in as well as four types of easy-fit tines for forward or reverse digging. The Crunchtiller was virtually identical to the earlier Monrotiller Minor and, after changing the cultivating rotor with a single driving wheel, it was used with a rotary scythe, toolbar, trailer and other attachments.

Mechanised Gardening, which also traded as Mechgard Ltd and was still selling the Monrotiller, Crunchtiller and Rototillers in the early 1970s, introduced four new Monro garden cultivators in 1974. The smallest 7½ hp four-stroke RO 320 had four forward and four reverse gears and an easily adjusted cultivating rotor with four working widths up to 31 in. The RO 40 was similar but with a 9 hp engine and the rotor width was adjustable from 20 to 36 in. The diesel-powered 10 hp RO 46 and 12 hp RO 66 also came with four forward and four reverse gears and an adjustable 24 to 48 in wide rotor and could be supplied with an optional two-speed rotor for digging to a maximum depth of 24 in.

The Monrotiller III was still available in 1977 along with the Terratiller, a very similar machine to the earlier Crunchtiller. There was a choice of a 3½ hp Aspera or a 4½ hp Clinton four-stroke petrol engine for the Terratiller which had a dry-plate clutch and an enclosed gear drive to the digging rotor. It was used for forward or reverse digging with pick tines, hoe tines, spring tines or slasher blades on the 10 in rotor extendable to a maximum 42 in working width.

Optional equipment for the TerraTiller included traction wheels for the rotor shaft, toolbar equipment, a small trailer and a tvo conversion kit for the 4½ hp Clinton engine. The 1974 Mechgard price list also included a range of Simar Rototillers with 7½ hp petrol to 12 hp diesel engines and various types of tine and working widths. The same garden cultivators were still advertised in 1981.

MOTOHAK

The Motohak, a wheel-less rotary cultivator made by Klien Motoren in Germany and imported by Minok Engineering Co of Maidstone was one of an increasing number of imported garden cultivators to appear on the British market in the late 1950s.

Power from the two-stroke Solo engine was transmitted through a disc clutch, a two-speed gearbox and chain drive to the rotor shaft. The rotor was made up with a series of cultivating stars and up to six could be used on each side of the rotor shaft. Wheels were not required for cultivating but pneumatic-tyred wheels were available for transport work. The throttle, clutch lever and twist grip gear change were placed on the adjustable handlebars.

In 1960 a strengthened version of the Motohak with an improved engine air-cooling system was launched at the Smithfield Show. Publicity material suggested that the 3.8 hp Motohak was a most versatile, powerful rotary cultivator that was easy to manoeuvre. The 98 cc two-stroke Motohak was still available in the UK in 1963 when Minok Engineering announced that the price had been reduced to £67 for the basic machine and a pair of cultivating tines was £3 10s 0d

MOTOM

Trojan Agricultural Sales of Croydon imported the Italian Motom M2 rotary garden cultivator in the early 1960s. Powered by a 51 cc air-cooled four stroke petrol/tvo engine the Motom had a three-speed gearbox and a power take-off shaft used for a sprayer or a hedge cutter. With a pair of steel or pneumatic wheels on the rotor shaft the Motom could be used as a small tractor with a conventional or reversible plough, a disc harrow, spike tooth harrow, cultivator and a garden truck.

MOUNTFIELD

G D Mountfield of Maidenhead, and later of Plympton in Devon, made the first Mountfield lawn mowers and M1 garden cultivators in 1962. Described as the ultimate in powered garden mechanisation, the Mountfield M1 cultivator was said to be the most efficient and comprehensive garden

1.94. The Motohak was made in Germany.

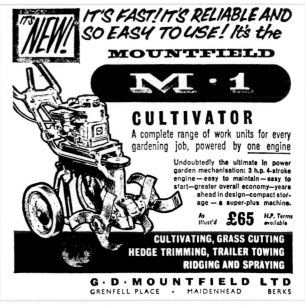

1.95. The first Mountfield M1 rotary cultivators were made in 1962.

1.96. The Mountfield M1 Estate rotary cultivator cost £187.50 in 1975.

machine ever devised and gave faster and deeper cultivation than any of its competitors. The 3 hp M1, which cost £65, had a four-stroke Kirby engine with a wind-up push button starter and a hand lever clutch to engage a twin vee-belt drive to the digging rotor. Attachments for the M1 included a rotary cultivator, grass cutter, hedge trimmer and a trailer.

The 1975 Mountfield price list included three models of the single-forward speed Mountfield M1 rotary cultivator, the 5 hp Monarch, 4 hp Estate and 3½ hp Super. With Briggs & Stratton four-stroke engines and two pairs of cultivating rotors the models cost £197.50, £187.50 and £168.75 respectively. Optional attachments for the three machines included extension rotors, a pair of driving wheels for use with a toolbar with hoe blades, a ridging body or cultivator tines and a small trailer. The M1 could also be used as a hand-propelled grass mower by attaching the engine and handlebar

1.97. The Mountfield M1 Gardener could be used as a cultivator or rotary mower.

assembly to an 18 in cut rotary mower deck driven by the same vertical drive shaft used for the cultivating rotor.

The Mountfield M1 Gardener with a 3½ hp Briggs & Stratton four stroke engine introduced in the early 1980s was also a dual-purpose machine which could be used with a rotary cultivator or an 18 in cut rotary lawn mower. The engine and handlebar assembly could be removed from the cultivator in less than a minute by loosening a hand wheel and lifting it on to the mower or vice-versa.

When G D Mountfield became part of the Ransomes Consumer Division in 1985, it continued

1.98. Mountfield Manor rotary cultivators were made in the early 1990s.

1.99. Norlett rotary cultivators were made in Norway.

the production of Mountfield rotary cultivators at the Plympton factory.

There were three models of the Mountfield Manor rotary cultivator in the early 1990s. The 3 hp Manor 3, the 5 hp Manor 5 and the Manor 5R all had Briggs & Stratton four-stroke engines with a vee-belt drive to the gear case and chain drive to the 24 in wide cultivating rotor. They all had fully adjustable folding handlebars but the Manor 3 was equipped with a front mounted-transport wheel which was folded up when the cultivator was in use. The Manor 5 had a pair of folding transport wheels at the rear and, unlike the single forward speed Manor 3 and Manor 5, the more expensive 5 hp Manor 5R had a reverse gear.

NORLETT

The Scandinavian-built Norlett TillerMate range of garden cultivators was marketed in the UK by Norlett Ltd, originally of Great Milton near Oxford but later of Dormer Road in Thame. There were three

models: the 3 hp TillerMate 3 and 3S, the 4 hp TillerMate 4S and the 5 hp TillerMate 5S. They all had Briggs & Stratton four-stroke engines, a 32 in wide cultivating rotor, a front bumper bar and rear transport wheel.

The TillerMate 3 was the most basic model with fixed handlebars and was meant for tilling kitchen gardens and allotments. The 3S was the de luxe version with adjustable handlebars and a front bumper bar. Except for its 4 hp engine the 4S had the same specification as the TillerMate 3S and the TillerMate 5S with a two forward and one reverse speed gearbox was recommended for use with a wheel drive attachment in the larger garden.

The TillerMate 4 and 5 Automatics were the most advanced models in the range with five forward speeds controlled at the touch of a hand lever. The 5 Automatic also had a reverse gear. A wide range of accessories, better suited to the more powerful models and used with a pair of pneumatic-tyred wheels on the cultivator rotor shaft, included a rear tool frame for cultivator tines, a ridging body, a weeder and a potato lifter. A plough, a trailer, a cutter bar and a dozer blade were also made for Norlett cultivators.

J T Lowe Ltd, which made hand-pushed Jalo hoes, cultivators and drills in the 1950s, imported Norlett 3 and 5 hp rotary cultivators in the mid-1980s. The 3 hp Norlett cultivator was a single-speed machine, the 5 hp model with a Briggs & Stratton engine had one forward and one reverse gear and a two forward and one reverse gearbox was fitted to the 5 hp Kawasaki-engined cultivator. The Norlett 6000 with a 5 hp Briggs & Stratton engine and a two forward and one reverse transmission was listed in 1988.

PEGSON

Made in the late 1940s by Geo Monro Ltd at Waltham Cross and in the early 1950s by Pegson of Coalville in Leicestershire, the Pegson Uni-Tractor had a 1.2 hp Villiers engine mounted within its single steel or pneumatic-tyred wheel running on eight steel guide rollers. The Pegson had an idler pulley to engage the vee-belt drive to the land wheel through two forward gears and an internal reduction gear.

The four-stroke Villiers fan-cooled petrol engine was mechanically governed and started with a rope wrapped around the crankshaft pulley. The six-inch wide wheel was steered with adjustable handlebars that could be offset to either side of the machine. Attachments included front and rear toolbars with cultivator tines, hoe blades and a ridging body. The Pegson Uni-Tractor cost £38 in 1951, an 18 in wide toolbar was 8s 6d and a complete set of toolbar accessories cost £11 7s 6d.

PLANET

Two-wheel Planet Junior garden tractors made by S E Allen & Co of Philadelphia in America were sold in Great Britain during the 1940s. The Planet A 1 and B 1 had 1 and 1½ hp Briggs and Stratton engines respectively, while their big brothers the HT and B HT were sold with either a 3 hp or 3½ hp power unit.

1.100. The Pegson Uni-tractor was made in the early 1950s.

1.101. Left – The Planet JR B-1; Right – The Planet JR 3 hp garden tractor.

They had a chain drive to the pneumatic-tyred wheels and the track width on the larger models could be adjusted from 14½ to 20 in.

A gardening magazine at the time advised readers that the Planet Junior garden tractor was well worth consideration and should be given a trial.

QUALCAST

The Derwent Foundry at Derby, which made its first lawn mowers in 1920, changed its name to Qualcast when it became a public company in 1928. Qualcast acquired the Suffolk Iron Foundry in 1958 and Charles H Pugh, makers of Atco lawn mowers, was taken over in 1965.

The first Qualcast garden cultivators were made in 1977. The gold and black Cultimatic Super and Cultimatic de luxe garden cultivators were similar to the Merry Tiller with the engine over the cultivating rotor. Accessories included a pair of rubber-tyred driving wheels for the rotor shaft, a small plough and a toolbar with hoe blades, cultivator tines and other implements.

The Cultimatic Super with a 98 cc four-stroke Suffolk engine, a two-speed drive mechanism with standard speed and overdrive and two depth control skids cost £149.95. The 4 hp de luxe Cultimatic with a Briggs & Stratton engine and reverse drive and twin swing-up transport wheels was £199.95.

1.102. The Qualcast Cultimatic Super had a 98 cc Suffolk engine.

The Qualcast Cultimatics became the 3 hp Atco cultivator and 5 hp Atco Cultivator de luxe in 1982. Apart from the addition of twin transport wheels on the 3 hp model and a new green and black colour scheme they were really re-badged versions of the earlier gold and black Qualcast machines. In 1985 the Qualcast Cultimatic B66 superseded the Atco Cultivator and Cultivator de luxe. The new green B66 Cultimatic with a 114 cc Suffolk engine cost £309.99 though the price had risen to £321.49 by the time it was discontinued in 1988.

REYROTO

1.103. The Reyroto was made in the early 1960s.

English Powerspades (Cultivators) Ltd, based at Over Wallop in Hampshire, exhibited a new range of Reyroto cultivators and miniature tractors at the 1960 Smithfield Show. The show catalogue listed the Silverspade Minor and Major rotary cultivators with two- or four-stroke engines.

Many attachments were available for the Reyroto, which was described as extremely manoeuvrable, light to handle and an efficient aid to gardening. The Reyroto Miniature Tractor, also available with a two- or four-stroke engine, was primarily designed for use with Jalo push hoe tools but it could also be converted to a rotary cultivator or used to tow a small trailer or dumper wheelbarrow.

ROWTRAC

1.104. Geo Monro sold the first Rowtrac garden tractors in 1932.

William Catchpole, a Suffolk farmer, designed the Rowtrac in the late 1920s and the first tractors were assembled for him by Robert Boby Ltd, a company making grain handling equipment in Bury St. Edmunds. The Rowtrac patent rights were sold to Geo Monro Ltd of Waltham Cross in the early 1930s and William Catchpole used the proceeds to help fund the development of the first Catchpole sugar beet harvester introduced in 1939.

Geo Monro became the sole distributor for the British-made two-wheeled Rowtrac motor cultivator which was exhibited at the 1932 Smithfield Show. Sales literature explained that growers' views on the design had been considered and a few machines manufactured by a famous British engineering firm had been loaned out for trial.

The Rowtrac specification included a 3½ hp Villiers 343 cc two-stroke engine with a Villiers flywheel magneto and a Villiers carburettor. The engine was cooled by fan blades on the flywheel and was started with the aid of a decompression valve by inserting the crank handle between two spokes in the right-hand wheel to engage with a starting dog on a shaft from the gearbox. Engine power was transmitted through a cone clutch to a single-speed constant mesh gearbox with a top speed of 3½ mph. The clutch and throttle levers were mounted on the handlebars, and ratchets in the wheel hubs made the Rowtrac easier to steer.

The Rowtrac 5 two-wheeled garden tractor on spade lug wheels was introduced in 1946 and manufactured for ten years. Advertisements

1.105. Early Rowtrac 5 models had a 3½ hp two-stroke engine.

explained that the '5' related to the Rowtrac's ability to carry out the five tasks of ploughing, discing, cultivating, hoeing and harrowing. The specification included a 3½ hp two-stroke Villiers engine with a flywheel magneto, Villiers carburettor and an air filter with metallic shavings. A cone clutch engaged the drive to the single-speed transmission and optional ratchet-type hubs with hardened pawls for power steering were controlled with a hand lever.

Sales literature explained that the weight of the tractor was concentrated over the

1.106. From 1951 the Rowtrac 5 had a 5 hp Villiers two-stroke engine.

wheels so that there was no waste of engine power. The Rowtrac was said to have a daily fuel and maintenance cost of about 5s 8d, based on ploughing an acre or hoeing between eight and ten acres in an eight-hour day on 2 to 2½ gallons of fuel.

Attachments for the Rowtrac included Ransomes general-purpose and digger ploughs, a disc harrow, cultivator, spike-tooth harrow, mower, potato lifter and a trailing seat. Pneumatic-tyred wheels, power take-off and a belt pulley were optional extras.

A more powerful Rowtrac 5 with a 5 hp two-stroke Villiers engine superseded the 3½ hp model in 1951 when it cost £127 15s 0d on standard pneumatic tyres or £116 on steel wheels. Otherwise the specification was very similar to the 3½ hp model but the daily fuel and maintenance cost had risen to about 6s 6d.

Accessories for the Rowtrac included a two conversion kit for the engine, a sprayer and a soil shredder.

1.107. There were three models of the Shay Rotogardener in the early 1960s.

SHAY

In 1952 J E Shay of Basingstoke acquired Power Specialities at Slough, makers of the Rotoscythe, the world's first rotary lawn mower.

Rotoscythe production continued at Basingstoke and the Shay Rotogardener 120 rotary cultivator, introduced in 1954, had the same 120 cc air-cooled two stroke 1¾ hp engine as the Rotoscythe. An over-centre jockey pulley served as a clutch to engage the vee-belt drive from the engine to an oil-immersed worm gear used to drive the front 16 in wide cultivating rotor, which with extra blades could be extended to 24 in. The half-gallon petrol tank, an integral part of the handlebars, held enough fuel and oil mixture for about three hours' work.

Easily handled in the kitchen garden, the Rotogardener 120 with a maximum digging depth of 8 in and with a pair of driving wheels on the rotor shaft, could be used with a rotary grass cutter, to haul a small trailer and operate spraying equipment. Other attachments included a toolbar with tines and

hoe blades, a generator and a flexible drive shaft for hand-operated Tarpen tools. It cost £49 10s 0d in 1954, a pair of rubber-tyred driving wheels for the rotor shaft adding £2 18s 6d to the price.

The Shay Rotogardener 80 with a JAP 80 cc two stroke engine, a front cultivating rotor and handlebars that could be offset to either side of the machine cost £44 when it was introduced in 1959. The Shay Rotogardener 125 with a 120 cc four-stroke Aspera engine was added in the early 1960s. Attachments for the 80 and 125 were similar to those available for the Rotogardener 120.

SIMAR

K V von Meyenburg was granted a patent in 1912 for his mechanical tillage machine with rotating spring-mounted tines and SIMAR, the Société Industrielle de Machines Agricoles Rotatives of Geneva, was one of four companies making ride-on tilling machines under the von Meyenburg patent.

The first of many versions of the pedestrian-controlled Simar Rototiller appeared in 1919 when Simar introduced the Rototiller 10 with a 5 hp water-cooled engine. The smaller Simar Rototiller 5 with a 5 hp air-cooled two-stroke engine, first made in 1927, was demonstrated to market gardeners in America. It was also the first mass-produced pedestrian-controlled cultivator.

Simar Rototillers, of Compton Street, London EC1 introduced its range to British growers in the mid-1920s. An advertisement in the May 1926 issue of the Implement and Machinery Review publicised a display of Rototillers at the Royal Dublin and the Bath and West agricultural shows.

Geo Monro Ltd at Waltham Cross in Hertfordshire were appointed Simar importers of the Rototiller 5 in 1929 when the Swiss company Piccard & Pictet also claimed to be the sole distributor of the Simar 5 in the UK. Geo Monro entered the Simar Rototiller 5 along with the Monro Monotrac and Duotrac garden tractors (page 61) in the World Agricultural Tractor Trials held in Oxfordshire in 1930.

Introduced in 1931, the Swiss-built Simar Rototiller 3, known as the Rototiller Junior, had a two-stroke 4 hp Simar petrol engine, a 14 in wide miller rotor with eight spring-mounted tines and a working depth of 8 to 10 in. Sales literature explained that the single-speed Rototiller 3 with an overall width of just 16 in had a forward speed of 5 furlongs (1,100 yd) an hour and used one gallon of fuel to do three-and-a-half hours of deep tillage work.

Although Geo Monro was the sole Simar distributor for the British Isles, A M Russell of the Grassmarket in Edinburgh, the service agents for Scotland, exhibited a range of Simar Rototillers at the 1931 Highland Show.

SIMAR
ROTOTILLER
Type 3

❦ The Patented Motor Rotary Soil Tilling Machine and Labour Saver for the Horticulturist and the Garden Owner.

❦ The greatest advance ever made in efficient preparation of the soil and cultivation of crops.

❦ The most marked evolution in the application of Power to Horticulture.

❦ The best modern substitute for the plough and the spade.

SOLE DISTRIBUTORS FOR THE BRITISH ISLES:

GEO. MONRO, LTD.,
Machinery Section,
WALTHAM CROSS, HERTS.

1.108. The Simar Rototiller Type 3 was introduced in 1931.

The Simar Rototiller 30, also known as the Junior, replaced the Rototiller 3 in 1937 when it was advertised as the Rolls Royce of motor cultivators. It had a 4 hp two-stroke Simar engine with a Lucas magneto, Amal carburettor and the fuel tank held four gallons of the required 16:1 mixture of petrol and oil. A special strap with a peg, which had to be

1.109. The Simar Rototiller 30 was also known as the Simar Junior. (Roger Smith)

1.110. The Simar 56A had an 8 hp two-stroke engine. (Roger Smith)

inserted into a hole in the starting pulley, was used to start the engine. Failure to locate the peg in this hole could result in the strap flying off the pulley, sometimes with painful results.

A dog clutch engaged drive to the two-speed gearbox and pins in the wheel hubs were used to engage high or low ratio or to disconnect the drive in order to allow the machine to be pushed by hand. Spring tines, knife tines or special hooked tines for deep tillage work were made for the miller rotor. Complete with a 10 or 15 in wide miller rotor, the Rototiller 30 cost £89 in 1937.

The 8 hp Rototiller 50 superseded the Rototiller 5 in the mid-1930s. Apart from its more powerful engine the specification was very similar to that of the previous model. A 24 in wide miller rotor was standard but optional 17 or 30 in wide rotors were available for the Rototiller 50. A later model with a 20 in wide miller rotor or narrow 15 in rotor cost £120. The Rototiller 51, made during the same period as the Rototiller 30 and 50, had the added advantage of a reverse gear.

Weighing 9¾ cwt the Rototiller 90, first made in the late 1930s, had a single-cylinder two-stroke engine using a petrol/oil mixture but with slight adjustments could run on petrol or diesel. A 40 in wide miller rotor was standard on the two forward and two reverse speed Rototiller 90 which used less that a gallon of fuel an hour when working at full load. Optional 33 and 51 in wide miller rotors were available for the Simar 90 which took three hours in low gear to cultivate an acre of land 14 in deep.

E C Geiger of North Wales in Pennsylvania was still selling the Swiss-built Simar Rototiller 56A, 35C, 21C and 20C to American farmers and growers in the mid-1950s. The 56A and 35C had the same specifications as the British-built machines but the 20C and lower speed 21C were single-gear cultivators with no reverse and the choice of an 8, 13 or 18 in miller rotor. Geo Monro Ltd discontinued British Rototiller 35 in 1958 and the last British Rototillers 56 and 56A were made on 1 April 1961.

SOLO

The Solo Combi 533 cultivator, introduced by Solo Power Equipment Ltd of Dudley, Worcestershire in 1971 had a 98 cc Solo single-cylinder two-stroke

1.111. The Solo Combi cultivator.

engine which developed 5 hp at 3,600 rpm. The specification included a centrifugal clutch, reduction gearbox and cultivating widths of 10, 16 or 24 in.

Ancillary equipment included a ridging body, a cutter bar mower, a rotary mower, a water pump and a generator.

TEAGLE

When W T Teagle of Blackwater in Cornwall introduced the single-wheel Jetiller garden cultivator to the gardening public in 1956 it cost £42 10s 0d complete with a toolbar, a set of spanners and a screwdriver. The Jetiller's 50 cc Teagle two-stroke engine with an automatic recoil starter was already in use with the Teagle Jetcut hedge trimmer and the clip-on Teagle Cycle Motor for bicycles. The engine was mounted on a sliding bed plate and power was transmitted by vee-belt and reduction unit to the pneumatic-tyred land wheel.

The prototype Jetiller had a chain reduction drive but an enclosed two-speed oil bath gearbox was used

1.112. A 50 cc Teagle two-stroke engine was used for early Teagle Jetiller garden tractors. (Teagle)

for production models. The engine bed plate was moved backwards and forwards with a hand lever to engage and disengage the vee-belt drive to the gearbox. The Jetiller was well suited to rowcrop work and a full range of Planet push hoe tools could be used on its front and rear toolbars.

In response to customer demand a new and more powerful 126 cc four-stroke Teagle engine replaced the 50 cc two-stroke power unit in 1959. The inlet and exhaust ports were close together and the long

inlet port served as a vaporiser which made it possible to run the engine on tvo.

The four-stroke engine added £12 10s 0d to the price but with its extra power the Jetiller could be used with a wider range of implements, including a small rotary cultivator, a potato digger and a flexible drive shaft for a hedge trimmer and other power tools. It cost as much to make the four-stroke Teagle engine as it did to buy single-cylinder engines from America and so from 1965 Briggs & Stratton engines were used for the Jetiller.

The Teagle Digoe, which cost £45 in 1958, was a 12 in wide rotary cultivator which, according to sales literature, could be converted to a two-wheel tractor in a matter of seconds. A vee-belt and small gearbox transmitted power from a 50 cc Teagle two-stroke engine mounted on the handlebars.

It was only necessary to undo one nut to detach the engine and handlebars from the rotary cultivator and fit them to a narrow-wheeled toolbar used with a selection of Planet-type cultivator tines and hoe blades or a cylinder lawn mower. A flexible power take-off shaft was used to drive a circular saw and a hedge cutter.

1.113. The Teagle Jetiller with a 126 cc four-stroke Teagle engine.

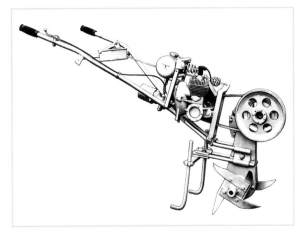

1.114. The Teagle Digoe was made in the late 1950s. (Teagle)

TEXAS

Chainsaw manufacturer Danarm of Stroud, Gloucestershire imported Texas cultivators from Denmark in the early 1970s. Similar in design to the Merry Tiller, the Texas de luxe cultivator with a 3 hp Briggs & Stratton engine, two sets of slasher blades and twin adjustable rear wheels cost £96 in 1972. By the late 1970s Danarm was importing four models of Texas de luxe cultivators with 3 or 5 hp Briggs & Stratton engines, while alternative Kawasaki engines were also used in the early 1980s.

The range started with the 3 hp, single-speed TV3 for use on the average-sized garden. The TD5 was similar but with a 5 hp engine, the 5 hp de luxe TD5B had one forward and one reverse gear and the TD5BR cultivator, also rated at 5 hp, had two forward speeds and one in reverse. A toolbar, cylinder and cutter bar mowers, a rotary brush and a lawn rake were among the attachments made for the Texas TD cultivator.

The mid-1980s Texas cultivator range included the 3 hp TV3 with a Kawasaki engine, the TV5, the TD5, the TD5B and the TD5BR, all with 5 hp Briggs & Stratton power units. The newer 5 hp 521 and 7½ hp 721 cultivators had Kawasaki engines and a two forward and two reverse gearbox. The 3 hp Texas Lilli 310 with a Briggs & Stratton engine and one forward speed was included in the list of Danish cultivators imported by Danarm in 1988.

A full range of attachments for all models of Texas

1.115. The Texas TD5 cultivator was made in Denmark.

cultivators included a plough, various tool frame accessories, a front cutter bar, cylinder lawn mower, tipper cart, sprayer and a generator-powered hedge trimmer.

Several models of Texas garden cultivator, including the Mini Tex 20, the Lilli, the Futura, and the Heavy Duty 521 special, were imported in the early 1990s by Stratford Power Garden Machinery of Stratford-on-Avon.

The Mini Tex with a two-stroke Tecumseh engine and 26 cm cultivating rotor was, with the handles folded, small enough to fit into the boot of a car. Briggs & Stratton engines were standard on the 3 hp Texas Lilli de luxe 320B and the 5 hp 520B, 542B and 562B but buyers could specify a Kawasaki, Tecumseh or Honda engine for their Texas Lilli.

a rotary sweeper and a 55 cm rotary mower were among the attachments for the Futura 2002.

TROY

Troy Agricultural Utilities of Hampton Wick, Surrey made the Mk I Troy Tractivator in the late 1940s. It had a 1½ hp JAP four stroke engine with a double vee-belt pulley and reduction gears giving top forward speeds of 1¼ or 2½ mph. It also had a power take-off, belt pulley and the choice of 14 in diameter pneumatic-tyred or steel wheels. The Tractivator, which used five pints of petrol in an eight-hour day could hoe, disc, drill, ridge, mow and spray.

Troy Agricultural Utilities, based at Surbiton in 1952, made the Mk I and Mk II Tractivators at its factory in Gloucestershire. The 1½ hp Villiers-engined Mk II had larger pneumatic-tyred or cleated steel wheels than the Mk I and attachments for both models included a plough, a harrow, a cultivator, a ridging body, a seed drill, a potato lifter, a sprayer and a trailer.

Another attachment for the Tractivator was a front-mounted rotary scythe with the mower disc interchangeable with a horizontal circular saw blade. The 18 in cut rotary scythe blade was suitable for cutting grass, kale or bracken and the 12 in saw blade could cut timber of up to 4 in diameter. It was suggested that when using pneumatic tyres the Tractivator's drawbar pull would noticeably increase if the optional bolt-on front weight box was filled with up to 100 lb of concrete or other heavy material.

Sales literature claimed that the Troy Tractivator was the world's foremost light tractor and cultivator and that in price, performance, versatility and reliability the 1½ hp Troy had no equal. It was pointed out that economically the Tractivator was just the job to do all work on a 1 to 20 acre holding, as its operation and maintenance did not require skilled labour and with an implement for every job all year round it would never need to be idle.

There were three models of Troy garden tractor in the late 1950s when the Spartan and Trojan replaced the Mk II Tractivator. All three had a JAP engine with the option of 14 or 20 in pneumatic tyres or 20 in cleated steel wheels and the wheel track was adjustable between 8 and 20 in.

1.116. . Optional 30 cm rotor extensions were made for Texas Lilli cultivators.

The single-speed Texas 320B and 520B had a 50 cm wide chain-driven cultivating rotor and a 55 cm rotor was standard on the one forward and one reverse gear Texas 542B and 562B. The two forward and one reverse gear Futura 2002 and the one forward and one reverse Futura 2000 and Futura 2001 had a 5 hp Briggs & Stratton engine and 60 cm heavy-duty cultivating rotor with an optional 30 cm extension.

A rake, weeder, ridging body, sulky seat and a cart were among the attachments available for the Futura 2000 and 2001. Right-handed and reversible ploughs,

1.117. The Mk I Troy Tractivator appeared in the late 1940s.

1.118. The Mk II Troy Tractivator.

The 1¾ hp Mk I and 2½ hp Spartan had a single-speed transmission while the 4 hp Trojan had a two-speed gearbox with a top speed of 5 mph. A wider range of attachments included general-purpose and digger ploughs, a rotary cultivator, onion lifter, turf cutter and a sulky seat.

TRUSTY

Tractors (London) Ltd, founded by a Mr J C Reach, assembled the first two wheel Trusty tractors at Tottenham in 1933 using parts made by fire extinguisher manufacturer Walter Kiddy & Co. Twenty-five Trustys were built during the next eighteen months but with no matching implements the tractor had to be used with modified horse equipment. This made it difficult to convince growers that the two-wheel tractor was just as good as a heavy horse although the machine did have the added advantage of not having to be fed when it was parked in its shed.

Tractors (London) Ltd moved to The White House at Bentley Heath near Barnet in 1938. Trusty was a household name in the world of two wheel tractors in the early 1940s when a wide range of Trusty implements including ploughs, cultivators, disc harrows, ridgers, hoes, mowers, trailers and transplanters was also made at Bentley Heath.

From time to time the Trusty tractor was mentioned in local newspapers. In 1946 a radio-controlled Trusty fitted with ex RAF radio equipment, a cylinder of compressed air to operate the controls and a single-furrow plough was

Details of the 1945
TRUSTY TRACTOR and IMPLEMENTS

1.119. A 1945 Trusty sales brochure.

1.120. A 1947 sales poster for Trusty tractors.

demonstrated to a large crowd. The event attracted wide press coverage and one magazine headline wondered if the era of armchair ploughing had arrived.

Later in the same year three radio-controlled Trusty tractors with single-furrow ploughs and a man at each headland to operate the controls and turn them round ploughed a field in competition with a Fordson tractor and a three-furrow plough. The ploughs all turned the same size furrows and when the work done by the three Trusty tractors was compared with that of the Fordson it was concluded that there was little to choose between them. Mr Reach commented that he was convinced that it would not be long before this new system of allowing the tractor to plough by itself would be in general use.

Trusty tractors hit the headlines again in 1948 when, in an attempt to increase food production, it was decided to plough up the wide verges on each side of the Barnet bypass. A convoy of five Trusty tractors ploughed the verges, barley was sown and

the sight of a tractor and binder at harvest time attracted still more press publicity.

Various petrol engines including Blackburn, Coborn, Briggs & Stratton and Wisconsin were used in the early days when many Trusty tractors left the factory with any engine that could be bolted to the tractor chassis and was available at the time. Tractors (London) Ltd made the 10,000th Trusty on 22 July 1947. There was a weekly output of 80 to 100 tractors at Bentley Heath during the peak production period in the early 1950s when about 100 people were employed there to meet orders from home and abroad.

The 1947 Trusty price list showed the Model 5 with a 5 hp JAP four stroke engine at £125 on steel wheels and the Model 6 with a 7½ hp Douglas four stroke engine or a 14½ hp Norton engine was £130, pneumatic tyres adding £10 to the price.

The comprehensive Trusty catalogue included every imaginable accessory from a one-way plough costing £30 and a flat roller costing £22 10s 0d to the Trusty transplanter with an easy-feed attachment priced at £45. Other items included an engine cover

1.121. Although most Trusty tractors seen on rally fields have pneumatic tyres most of them left the factory on steel wheels. (Roger Smith)

1.122. A 1948 Trusty with a Douglas engine. (Roger Smith)

1.123. This apple green Trusty Mk 5 has a 5 hp JAP engine and a three forward and one reverse gearbox. (Roger Smith)

for £2, a large grease gun for 7s 6d and a 10 gallon drum of Trusty engine oil for £3 15s 10d.

Although some late 1940s and early 1950s Trusty tractors had a Villiers Mk 40 engine, most were powered by a single-cylinder side-valve Douglas, JAP or Norton. A Petter diesel engine was used for some of the later Model 6 tractors but when it was found to be too heavy it was replaced with a Sachs diesel engine which suffered from starting problems. Although short on styling, the Trusty tractor was an extremely functional machine with its automatic centrifugal clutch, chain drive to a countershaft and a roller chain to drive to both wheels. Dog clutches in wheel hubs provided power turning at the headland. With no gearbox on early models the Trusty had a top speed of 2 mph.

It was reported in a July 1944 machinery journal that Tractors (London) had been experimenting with a reverse drive, which it planned to introduce in the foreseeable future. It was explained that the reverse drive mechanism consisted of an epicyclic

unit and a reversing flywheel mounted on the offside of the tractor. With the engine at full throttle a lever was used to engage drive to the reversing flywheel. This slowed the tractor, allowing the epicyclic unit to engage reverse drive automatically.

An optional factory-fitted reverse gear introduced in 1948 overcame the need to pull the tractor backward from a corner manually. Marketed as a 'Safety Reverse' it consisted of an externally mounted aluminium gearbox with a spring-loaded lever to change from forward speed to reverse. The reverse unit, which cost £10, was also offered to owners of older Trusty tractors.

The spring-loaded lever served as a deadman's handle and reverse was automatically disengaged when the hand lever was released to reduce the risk of the operator stumbling or falling when backing the tractor. The 5 hp Trusty 5 and 6 hp Trusty 6 gained a three forward and one reverse speed Albion gearbox in 1951 but fewer than 300 had been made when the Trusty tractor was discontinued in 1967.

Trusty implements were hitched to a swinging drawbar or wheeled toolbar beneath the tractor's unusually long handlebars and a two-wheel bogey seat was available for use with a cutter bar mower, transplanter, trailer and other implements, and a single-wheeled seat was made for use with the Trusty Greyhound plough. A self-lift mechanism used with soil-working implements was engaged by pushing on the lift handle which used the resulting torque reaction to lift the toolbar from work. A power take-off shaft was standard but the belt pulley was an optional extra.

The Trusty Imp, introduced in 1949, was a scaled-down version of the original Trusty with a 2½ hp Villiers Mk 25 four stroke engine, a centrifugal clutch and simple chain drive to the wheels. The Imp, complete with an 8 in

digger plough and 3 ft wide light cultivator, cost £100. A 1950 sales leaflet claimed that the Imp led the field with its ample reserve of the right type of power, its silky start, smooth running, self steering and a purpose-designed tool for every job on the holding.

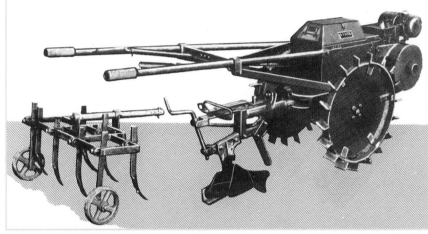

1.124. The Trusty Imp was introduced in 1949.

1.125. An optional flexible drive shaft for the Trusty Earthquake could be used with a hedge trimmer, a chain saw and a hand-held rotary cultivator. (Roger Smith)

1.126. The Trusty Whirlwind was a combined hoe and grass cutter.

The Trusty Earthquake rotary cultivator with a wide range of attachments was introduced in 1961. Buyers could choose a 2½ hp Mk 20, 3 hp Mk 25 or a 4½ hp Mk 40 Villiers air-cooled power unit with a dry plate clutch and three-speed Albion gearbox with an optional reverse gear.

Later models of the Earthquake were supplied with a 3 or 7½ hp Villiers engine. The Earthquake had individual chain drive and reduction gearboxes for both wheels and a heavy-duty roller chain running in an oil bath was used to drive the three-speed cultivating rotor. Its maximum working depth was 7 in. The 2½ and 3 hp versions were equipped with a 14 in wide cultivating rotor; the 16 in wide rotor on

the 4½ hp model was controlled by a separate dog clutch.

Prices were £125, £140 and £184 respectively and there was a full range of attachments including a plough, cultivator, ridging bodies and a 5 cwt trailer. The Earthquake had a power take-off with a belt pulley and a flexible drive for a saw or hedge trimmer and could also be used with a front-mounted cylinder mower, cutter bar mower or a rotary mower.

Tractors (London) also made the Trusty Mowmotor, the Trusty Whirlwind combined hoe and grass cutter and the Trusty Weedsweeper. The mid-1950s Mowmotor with a ⅓ hp two-stroke engine could be clamped to most makes of hand-pushed side-wheel and roller mowers. The weight of the engine unit held its chain-driven friction roller firmly against one wheel of a side-wheel mower or the driving roller on a hand mower. Drive was engaged or disengaged by lowering or lifting the Mowmotor engine with a hand lever.

The Weedsweeper and Whirlwind, each with a ⅓ hp engine, were introduced in 1957. A vee-belt drive was used for the Weedsweeper's 6 in rotary hoe and Whirlwind combined hoe and grass cutter. The Whirlwind was used to trim rough areas of grass with a small mower or remove weeds with a set of revolving pins on a hoeing disc.

WESTWOOD

Westwood Engineering made the Groundhog rotary cultivator at the Plympton factory in Devon in the 1970s. The economy model of the Groundhog G99 with a 3 hp Briggs & Stratton engine, a vee belt and roller chain drive to the cultivating rotor and fixed handlebars was recommended for the home gardener. The 3 hp G3 Groundhog was similar to the G99 and, like the 4 hp G4 and 5 hp G5, the handlebars could be moved to either side of the machine.

1.127. The Westwood Groundhog was made in the 1970s.

Groundhogs were equipped with tine guards and transport wheels, and the 26 in wide cultivating rotor was easily reduced to 14 in for working in rowcrops. Accessories included a toolbar, a ridging body and a choice of solid rubber, pneumatic-tyred or spade lug wheels.

The Westwood garden cultivator design was sold in 1980 to Qualcast which made the cultivator with some modifications exclusively for Flymo. The 3 and 5 hp Flymo engine-over-rotor cultivators with an orange and brown colour scheme were discontinued in 1984.

WOLSELEY

Clayton B Merry obtained an American patent in 1952 for his design of a walking cultivator. By the mid-1950s the Merry Manufacturing Co at Marysville in Washington was building the engine-over-rotor Merry Tiller garden cultivator. By the end of the decade there were four models of the Merry Tiller made for the American market, initially by the Merry Manufacturing Co but later by Merry Tiller Inc of Birmingham, Alabama.

Wolseley Engineering Ltd, formerly the Wolseley Sheep Shearing Co, which manufactured stationary engines, electric fencers, sheep shearing tackle and other equipment in the 1930s and 1940s, made the first Merry Tiller walking cultivators under licence in 1957. Publicity material explained that the Wolseley Merry Tiller's cultivating rotor propelled the machine while stirring the soil. With a pair of traction wheels on the rotor shaft it could be used for ridging, hoeing, mulching and mowing.

The Merry Tiller had a four-stroke 75 cc Suffolk Punch petrol engine with a vee-belt drive to the 12 in wide cultivating rotor which also propelled the

The Smallholder, April 28, 1962

MERRY TILLER

prepares seed beds

BETTER · FASTER · EASIER

Price: with Transport Wheels
less Rotors £56.15.0 or Easy
Terms.
All the power goes direct to the
rotors *where it counts!

the finest drive
principle in the
rotary cultivator field
gives MERRY TILLER
its tremendous
capacity for hard work

**WIDE CHOICE OF
ACCESSORIES MAKE
IT USEFUL EVERY
DAY OF THE
YEAR**

WOLSELEY

MERRY TILLER

1.128. The 1962 Wolseley Merry Tiller.

machine. Additional rotor blades were available to increase the working width to 24 in and a pair of wheels was provided for transport purposes.

Wolseley Engineering and Hughes Engineering merged in 1958 to form the Wolseley Hughes Group. Five years later the group was joined by the lawn mower manufacturer H C Webb Ltd which continued to trade as a separate concern and retained control of the Webb lawn mower business. H C Webb changed its name to Webb Lawnmowers Ltd in the late 1960s and within a few years a large part of the

Wolseley factory was devoted to lawn mower production.

With a growing demand for motor mowers the two companies integrated in 1973 to form Wolseley Webb. The Merry Tiller remained in production throughout this period and it was still being made when the Birmid Qualcast group acquired Wolseley Webb in 1984.

Merry Tillers were made with a 1 or 2½ hp engine in the late 1950s and with the standard width cultivating rotor they cost £57 and £59 respectively. A wide range of attachments used with a pair of pneumatic-tyred driving wheels on the rotor shaft included a toolbar, trailer, front cylinder lawn mower, cutter bar, saw bench and a flexible drive shaft for a hedge trimmer and sheep shearing equipment. The use of wide rubber-tyred traction rollers was recommended when mowing lawns with the Merry Tiller's cylinder mower.

The Merry Tiller Major with a 2½ hp Briggs & Stratton four stroke lightweight aluminium engine and the Professional with a heavy-duty cast-iron 3¼ hp Clinton engine were current in 1960. Both had a recoil starter, and a vee belt tensioner pulley served as a clutch to engage the roller chain drive to the rotor shaft. The handlebars on the Professional were adjustable for height and could be offset to either side of the machine. A reverse drive attachment using a second vee-belt and set of pulleys was an optional extra for the Major.

The 1965 Merry Tiller range included the Major, Professional and the Titan with Briggs & Stratton 4 or 5 hp engines. A two-speed vee-belt drive, tensioned by an over-centre jockey pulley, provided two rotor speeds. The Titan's cultivating rotor with a triple roller chain reduction drive from the engine could be used with standard or slasher tines.

rotor shaft for ridging, rowcrop work and transport. Steel fenders, designed to protect the operator and the machine from mud, dust and stones, were added to the list of optional extras for all models of the Merry Tiller in 1967.

The 5 hp Merry Tiller Titan GT - with a gear transmission - and the 2 hp Trident, both with Briggs & Stratton engines, were added in 1971. The Titan GT had a two forward and one reverse gearbox, which combined with a two-speed vee-belt drive, gave four forward speeds and two in reverse. Users were advised to use the two lower gears when rotary cultivating but all four speeds could be used for ploughing, cutter bar mowing and inter-row cultivating.

The Trident small garden cultivator with a three-tined vertical rotor on an extended shaft in front of the engine was a departure from the familiar engine-over-rotor Merry Tiller design. It also differed from the Merry Tiller in that the operator was required to walk backwards with the machine while sweeping the tined rotor through an arc in order to avoid leaving any footmarks on the freshly cultivated ground.

1.129. The Merry Tiller Major with a 2½ hp Briggs & Stratton engine was used with various rowcrop tools and attachments.

Attachments for the Titan included a toolbar with tines and hoe blades, a 5 cwt trailer and a belt-driven water pump. Sales literature explained that the Merry Tiller would do every job in the garden with its many attachments including an unusual reversible plough similar in shape to a ridging body with slatted mouldboards which turned a furrow from 6 to 10 in wide and deep.

The Wolseley Twin Six, introduced in the early 1960s, could be used with up to six pairs of heavy-duty slasher blades to give working widths from 3 to a massive 6 ft. A 9 hp Briggs & Stratton engine provided the power to drive the rotor through a vee belt drive and a two forward and one reverse gearbox. Pneumatic-tyred wheels were fitted on the

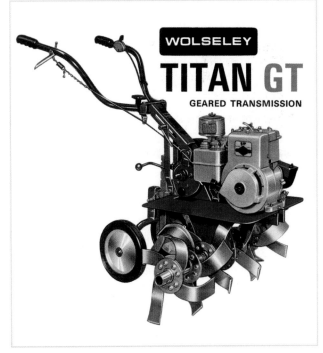

1.130. The Merry Tiller Titan could be converted from a rotary cultivator to an inter-row hoe in a matter of minutes.

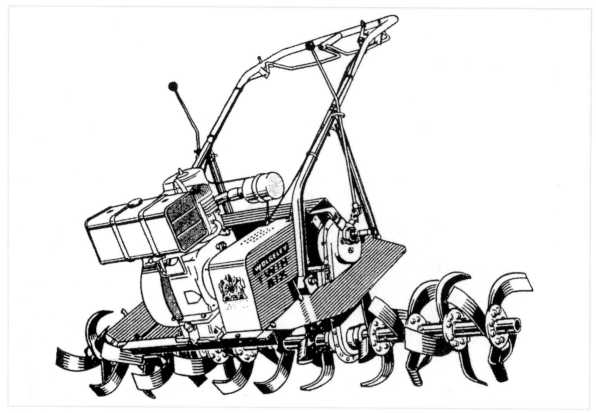

1.131. A maximum cultivating width of 6 ft was possible with the Wolseley Twin-Six.

The 1975 Wolseley Webb catalogue included the 3 hp Merry Tiller Major and Super Major, the two-speed 5 hp Titan and the 5 hp Titan GT with four forward and two reverse gears and the 9 hp Twin-Six walking tractor. Within a year the Merry Tiller Major had a new 4 hp engine and the Super Major was upgraded to 5 hp.

The Major cost £225 in 1976; the Super Major was £235, the Titan cost £295 and the Titan GT was £395. Six years later the Wolseley Webb Merry Tiller range, all with Briggs & Stratton power units, included the new 3 hp Cadet with roller chain transmission, the 4 hp Major and 5 hp Super Major with optional reverse drive and the range-topping 5 hp Titan and 7 hp Titan GT.

Reverse drive was standard on the Titan which had a triple chain reduction drive to the cultivating rotor with a maximum working width of 44 in. Sales literature explained that the four forward and two

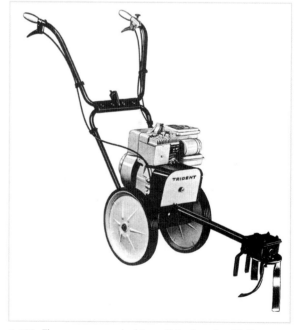

1.132. The user was required to walk backwards with the Wolseley Trident.

1.133. The 3 hp Wolseley Cadet appeared in the early 1980s.

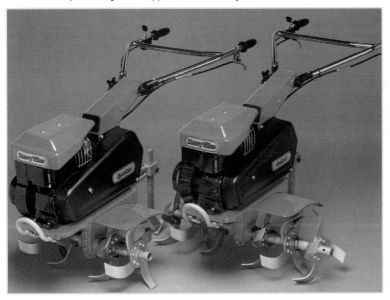

1.134. The Spartan and Centaur were added to the Merry Tiller range in 1983.

reverse gear Titan GT would dig up to 14 in deep with its 50 or 55 in wide cultivating rotor.

Two models of the Wolseley Webb Wizard garden cultivator were introduced after H C Webb joined the Wolseley Hughes group. The 3 and 5 hp Wizards had Briggs & Stratton engines to drive the slasher rotors with a maximum working depth of 12 in. Complete with 26 in wide slasher rotor and folding front-mounted transport wheel, similar to that on the Merry Tiller Cadet, the 3 hp Wizard cost £166.67 in 1978 while the 5 hp model was £190.67.

The new Spartan and Centaur Merry Tillers, along with the improved and re-styled 5 hp Super Major and Titan, the 7 hp Titan GT and the 3 hp Cadet, all still with Briggs & Stratton engines, were announced in 1983. The 5 hp Spartan and Centaur were single-speed machines with a vee-belt jockey pulley clutch, two-stage reduction chain drive, 24 in wide digging rotor and rear transport wheels.

Reverse drive was standard on the Super Major, Centaur and Titan, while with its four forward and two reverse speed rotor shaft driven by a combined vee belt and gear transmission system the Titan GT was said to be the ultimate in garden cultivators.

Merry Tiller production was transferred to the Suffolk Lawnmowers factory at Stowmarket in 1984 when Birmid-Qualcast acquired Wolseley Webb. The new owners gradually reduced the Merry Tiller range, with the Titan being discontinued in 1989 and the Major, Super Major and Cadet going out of production in 1990. The last Merry Tiller Titans were made in January 1991.

The Merry Tiller returned to the UK market in 2004 when Allen Power Equipment introduced four models made by MacKissic in America. The basic design has hardly changed from the original Merry Tiller with the cultivating rotor under the engine, a vee-belt clutching system and roller chain drive to the rotor tines.

The Tiny, with a 12 in wide rotor and a 2 hp two-stroke engine is the smallest MacKissic Merry Tiller. Four-stroke Briggs & Stratton engines with a recoil starter provide the power for the 4 hp Minnie and 5½ hp Suburban with working widths of 12 to 18 in and 18 to 26 in respectively.

The International with a 6½ hp Intek or 5½ hp Honda engine and a twin vee-belt drive combined

1.135. The 5½ hp Merry Tiller International was made by McKissic in America and sold in the UK by Allen Power Equipment.

with a jockey pulley clutch and triple reduction gear transmission is the largest member of the MacKissic mid-tine Merry Tiller family. Described as a rental duty tiller, the International could be used with a 14, 26 or 37 in wide tilling rotor with various types of tine.

Chapter 2
Ride-On Tractors

In the 1940s many smallholders and market gardeners needed to hire the services of a local farmer to plough their land. As most farm tractors were not suitable for working in closely spaced rowcrops this work required long hours walking behind two-wheel garden tractors. The introduction of more versatile three- and four-wheel 5 to 10 hp tractors such as the Byron, Garner, Gunsmith and Trusty Steed from the mid-1940s meant that this work could be carried out while riding on a tractor seat. Some growers used a Ransomes MG or a Bristol crawler, others tended their crops with a self-propelled toolbar.

However, the popularity of these small tractors was short lived, as they were not able to compete with the more sophisticated Ferguson TE 20 tractor which, complete with its fingertip control hydraulic system, cost just over £300. The Ferguson TE 20, Fordson Dexta and similar small farm tractors met the needs of smallholders for several years but as farm tractors became bigger and more powerful a new generation of small compact tractors was brought in from America, Europe and Japan to fill the gap.

ALLEN

At the 1961 Smithfield Show John Allen & Sons of Oxford, which later traded as Grove Allen Ltd and Allen Power Equipment Ltd, introduced the Allen Motostandard ride-on tractor built by Gutbrod in Germany. The mid-1960s Allen Motostandard had a 9 hp single-cylinder four-stroke petrol engine with electric starting. The specification included a four forward and two reverse gearbox,

diff-lock, foot brakes, hand brake and power take-off shafts at the front and rear.

The 10½ hp Motostandard 1031 and 7 hp 1016 with Briggs & Stratton engines were introduced to the UK by John Allen in 1967. The 1031 had a single-cylinder petrol engine with a 12 volt electric starter, a three forward and one reverse gearbox and a high/low ratio box doubled the range of gears. Other features included a diff-lock, a power take-off shaft and an adjustable wheel track. The smaller Motostandard 1016 was, apart from the lack of a high/low ratio box, very similar to the 1031.

A wide range of implements included ploughs and tillage implements and also left- and right-handed single-furrow mounted ploughs used alternately for one-way ploughing.

The Allen organisation introduced the new four-wheel Allen Motostandard Gutbrod Superior tractor

2.1. A mid-1970s Allen garden tractor.

91

2.2. The Allen Motostandard Gutbrod Superior was introduced to British growers in 1965.

to the UK in 1965. The 9 hp tractor had a MAG engine with Bosch starter generator, four forward and two reverse gears, diff-lock, independent brakes and a two-speed power take-off. An advertisement explained that although it was small in size the Gutbrod Superior was mighty in performance and capable of mechanising horticultural tasks in a way that had to be seen to be believed.

John Allen & Sons acquired the rights to manufacture and market Croft Mayfield two-wheel garden tractors from Mayfield Engineering Ltd in 1965. This arrangement continued until Arun Tractors bought the Croft Mayfield designs in 1975. The four-wheel Mayfield Merlin ride-on twin-blade mower with a rear towbar (page 252) was one of the products made during the Allen Mayfield era.

There were four models of Allen Motostandard garden tractor in the mid-1970s. The 1010 with an 8 hp Briggs & Stratton engine and the 1010A with a 7 hp Tecumseh power unit had electric starting, vee-

belt drive to a three forward and one reverse gearbox and a twin-blade 32 in wide rotary mower deck. Equipment for the 8 hp Motostandard 1017 with three forward gears and one in reverse included a single-furrow plough, spring tine cultivator and a spike tooth harrow together with rotary and cutter bar mowers.

The 12 hp Motostandard 1032 had a four forward and two reverse gearbox, a diff-lock, rear power take-off, an optional front power shaft and wheel track adjustable from 1ft 8 in to 2 ft 2 in. Implements for the 1032 included a plough, a rotary cultivator, a rotary brush, spraying equipment and rotary and cutter bar mowers.

A 14 hp Gutbrod MAG engine provided the power for the four forward and one reverse speed Motostandard 1050 with a diff-lock, front and rear power take-off shafts, a 32 in twin-blade rotary mower deck and a rear drawbar for towing a gang mower. The Allen Motostandard 1050D was the diesel-

2.3. The 18 hp Allen Roper 18T garden tractor had a twin-cylinder Onan petrol engine.

engined version of the 1050 with a 14 hp Farymann four-stroke power unit.

Grove Allen introduced the American-built Roper four-wheel tractors to the UK market in 1973 and Allen Power Equipment continued with this franchise until the early 1980s. There were four Allen Roper models in 1975. The smallest RT8R was an 8 hp garden tractor with three forward speeds and a recoil starter. The six forward speed RT8E had the same engine with an electric starter while the RT10E was similar to the RT8E but with an 8 hp power unit.

The Allen Roper RT13E garden tractor with a 13 hp Briggs & Stratton engine was available with a six-speed gearbox or hydrostatic transmission. A later version had eight forward and two reverse gears. All three Allen Roper machines could be used with various attachments, including an underslung mower deck, a sweeper, a shredder, a plough and a cultivator.

Three versions of the Allen Roper RT16E with the option of a single- or twin-cylinder 16 hp engine and a six-speed manual or a hydrostatic transmission were listed in 1975. Allen Roper ride-on tractors in the late 1970s included 11 and 14 hp models with electric starting and a six forward speed gearbox and the 16T with a Briggs & Stratton twin-cylinder petrol engine, eight forward and two reverse gears.

Roper yard and garden tractors, including 14, 16 and 18 hp Briggs & Stratton powered yard tractors and an 18 hp garden tractor with a Kohler engine, were marketed by Salopian Kenneth Hudson from Market Drayton in Shropshire in the mid-1980s. The 16 hp yard tractor had a hydrostatic transmission. A five- or six-speed manual gearbox was used on the other models, and all four had an electric starter and a power take-off shaft.

2.4. The Allis Chalmers Model G tool carrier.

ALLIS CHALMERS

The 10 bhp Allis Chalmers Model G self-propelled tool carrier introduced in 1948 was made at Gadsden, Alabama in America, although some were assembled in France and a few were sold in Great Britain and continental Europe. The Model G tool carrier, classed in America as a one 12 in plough tractor, had a four-cylinder side-valve Continental AN62 petrol engine with electric starting, a single-plate clutch and a four forward speed and a low reverse gearbox with a top speed of 7 mph. The gearbox, differential and final drive gears were mounted above the rear axle.

The Model G had independent rear wheel brakes and the wheel track was adjustable from 36 to 64 in. A belt pulley was offered as an optional extra but it did not have a power take-off shaft. A hand lever was used to raise and lower the front and underslung toolbars on early models and hydraulic linkage was added to the standard specification in the early 1950s. Just under 30,000 Allis Model G tractors had been made when production ceased in 1955.

ASHFIELD

The Ashfield 15.4E ride-on tractor, made by Ashfield Agricultural Products Ltd of Derby, was sold with a 9, 11 or 16 hp Lister diesel engine. The specification included a 12 volt electrical system, a four forward and one reverse gearbox, hydraulic linkage and power take-off. Implements for the Ashfield tractor included a front-end loader, plough and cultivation equipment, a transport box and a tipping trailer.

ATCO

Atco introduced a range of lawn and garden tractors in the late 1970s for the domestic market. Although designed as ride-on rotary mowers with a mid-mounted mower deck, they were supplied with a drawbar for towing a small trailer, a lawn sweeper, a mini gang mower and other trailed turf care equipment. The 7/32, 8/32, 8/36, and 11/36 model numbers of the Atco lawn tractors denoted the horsepower of the single-cylinder Briggs & Stratton air-cooled engines and cutting width of the rear-discharge mower decks.

The four lawn tractors had a three forward and one reverse gearbox, while the 8 and 11 hp tractors had electric starting. The grass clippings were returned to the ground and if required they could be collected with a lawn sweeper towed from the drawbar. The 16/42 garden tractor had a twin-cylinder 16 hp Briggs & Stratton engine, four forward gears and one reverse, electric starting and a side-discharge mower deck with three cutting rotors.

BARRUS

E P Barrus of Bicester was importing the MTD 990/16 hydrostatic tractor with a 16 hp Briggs & Stratton petrol engine in 1977. It had a direct drive from the engine to a dual-range hydrostatic transaxle and differential. A single lever provided infinitely variable speeds in forward and reverse with a top speed of 8 mph in both directions. Attachments for the MTD 990/16 which had disc brakes, power take-off and hydraulic lift included a rotary mower, cutter bar mower, plough, cultivator, disc harrow, rotary cultivator and a snow plough

BEAN

A small number of three-wheel Bean self-propelled toolbars, originally designed by a Mr Bean for his smallholding, was made at an aircraft factory in Blackburn in 1945. Humberside Agricultural Products of Brough in East Yorkshire took up the manufacture later in that year and the Bean was made there for about ten years. The tricycle-wheel Bean tractor had a rear-mounted Ford 8 industrial side-valve petrol engine, three forward gears and one reverse, a top speed of

2.5. The early 1980s 16 hp Atco 16/42E ride-on tractor with a 42 in cut rotary mower deck could also be used with a plough, cultivator and an engine-driven rotary cultivator.

2.6. The three-wheeled version of the Bean self-propelled Toolbar was awarded a Silver medal at the 1947 Royal Show.

4 mph and a Ford 10 cwt axle to drive the pneumatic-tyred rear wheels. Electric starting and independent brakes were standard equipment.

The driver sat in front of the engine with an unobstructed view of the work and the single front wheel was steered with a curved tiller handle. Implements for the Bean self-propelled toolbar, which cost £260 ex works in 1949, included cultivating tines and hoe blades, seeder units, a fertiliser spreader and a crop sprayer.

A four-wheel version of the Bean with tiller steering, a choice of four wheel track settings and the capacity to hoe four, five or six rows in one pass was added in 1948. The Bean cost £295 when it was exhibited at the 1950 Smithfield Show. The five-row hoe attachment was £45 and five Bean seeder units with large capacity hoppers added £65 to the price. Sales literature explained that the three-wheel Bean rowcrop tractor priced at £280 was still available for the discriminating grower who wished to carry out intensive cultivations on a smaller scale. It was also pointed out that the Bean was so easy to handle the operator would not be tired by teatime and would be able to work overtime without fatigue.

2.7. The four-wheel Bean self-propelled toolbar appeared in 1948.

2.8. The Strathallan Bean self-propelled power unit.

By the late 1950s Thomas Green & Son of Leeds had taken over the manufacture of the rowcrop tractor and its attachments. At this time the three-wheel tractor cost £340 and the four-wheel tractor £360.

The Strathallan Engineering & Construction Co of Auchterarder, Perthshire introduced a new version of the Bean self-propelled power unit in the mid-1960s. The three-wheel layout and tiller steering was retained for the new model. It had a 16 hp twin-cylinder Petter diesel engine with electric starting, a single-plate clutch, a three forward and one reverse gearbox and a totally enclosed oil-immersed rear axle. The Strathallan Bean, with a top speed of 6 mph, had inboard hydraulic brakes, hydraulic linkage and track width adjustment from 48 to 76 in.

Attachments included a mid-mounted toolbar with cultivator tines, hoe blades, weeder tines, a crop sprayer and a set of seeder units.

The tiller handle was replaced with a steering wheel on an improved mid-1970s version of the Strathallan Engineering Bean when there was a choice of a 16 or 24 hp Petter diesel engine. In 1979, when demonstrated with a six-row band spraying unit and two 200 litre saddle tanks at the National Spring Sugar Beet Demonstration in Lincolnshire, the price of the Strathallan Bean had risen to £3,695.

Strathallan Engineering was still building Bean self-propelled toolbars in 1981 but within a year or two it was being made by Bean Equipment in Ely, Cambridgeshire. A Mk III version of the four-wheel Bean Beaver toolbar cost £6,500 when it was announced in 1984. The driver sat alongside a centrally mounted 24 hp Lister twin-cylinder diesel engine and the specification included a twelve forward and two reverse gearbox, power steering and a hydraulically operated toolbar.

BMB PRESIDENT

In the 1930s British Motor Boats of London imported American-built marine engines and Simplicity two-wheel garden tractors to be sold in the UK as BMB garden tractors. The company moved to Banbury in 1940 and with Simplicity tractors in short supply Shillans Engineering Ltd, also of Banbury, was contracted to make a limited number of the small BMB Hoe-Mate garden tractors. Brockhouse Engineering bought British Motor Boats in the mid-1940s and transferred production of BMB garden tractors to Crossens near Southport in 1947 where the two-wheel Plow-Mate, Cult-Mate and Hoe-Mate were made until the mid-1950s.

In 1950 the four-wheel BMB President tractor was introduced to the farming public at the Royal Agricultural Show. It was in quantity production by the end of that year and was made for the next six years. Power was supplied by an 8/10 hp Morris four-cylinder petrol engine which could be converted to run on tvo. Sales literature claimed that when the Morris petrol engine was running at its rated speed of 2,500 rpm it developed 18½ bhp and 15 hp at the belt pulley.

The engine was equipped with a Solex updraught carburettor, a large capacity oil bath air cleaner and a thermo-syphon cooling system. The 6 volt electrical system included a generator, a starter motor, coil ignition and optional lights. The instruction book explained that a starting handle was provided just in case it might be needed. Later models of the BMB President had a 12 volt electrical system.

The transmission consisted of a single-plate dry

2.9. The BMB President tractor was made for six years.

clutch, a three forward and one reverse sliding-mesh gearbox, spur gear final drive and independent rear wheel brakes with a parking lock. The President had top forward speeds of 1¾, 3¾ and 8 mph and the wheel track was adjustable from 40 to 72 in. Sales literature explained that at maximum engine speed the rear wheels turned at 70½ rpm in top gear and at 25 rpm when in reverse. The basic price of the President in 1951 was £239 10s 0d complete with a swinging drawbar but this had increased to £276 10s 0d by 1953.

Optional equipment included a high-speed gearbox which gave a 20% speed increase in all gears, power take-off, belt pulley and hydraulic linkage or an alternative hand-lift toolbar. The hydraulic unit was bolted to the side of the transmission housing and gearbox oil was used for the mid-mounted toolbar and rear three-point linkage rams.

The mid-mounted toolbar had a simple depth-control system with depth-limiting pads, which rested on the front axle when the hoes and other toolbar equipment were in work. A screw-handle adjustment on the limiting pads controlled the depth of the hoe blades and when working on uneven surfaces the pivoting action of the front axle raised and lowered

either side of the toolbar in order to maintain the correct depth.

Publicity material pointed out that the tractor's wasp waist combined with its wire-mesh driving platform provided wonderful visibility for rowcrop work. It was also explained that exhaustive tests had revealed that with its extraordinary tyre adhesion and a terrific drawbar pull the President gave an astonishing performance out of all proportion to its size. Potential customers' attention was also drawn to the fact that the six-month warranty period did not apply to claims resulting from hiring out, racing, pace making or speed testing!

The President was one of the more successful light four-wheel tractors of the 1950s. Vineyard and orchard models were added in 1954 when the tractor, advertised as the Brockhouse President, cost £276 10s 0d. It was suggested that although there was a range of ploughs specially designed for the tractor it would do first-class work with practically any trailed plough.

However, the ceaseless demand for more engine power brought a general decline in the sales of small tractors and the last Brockhouse Presidents were made in 1956. The later Stockhold President based on the tractor was exhibited at the 1957 Smithfield Show (page 138).

BOLENS

The Bolens family business was founded in Port Washington, Wisconsin in 1850. W H Bolens, from a later generation of the family, bought the American tractor and engine manufacturer Gilson in 1914. The first Bolens garden tractors appeared in 1919 and, trading as Gilson Bolens, the company made its first lawn tractors in 1931.

Following another name change, this time to Bolens Products Co, the Standard and Hi-boy Bolens Husky Gardener two-wheel tractors appeared in the early 1940s. Immediately

2.10. The Bolens Ride-a-Matic was introduced to the UK market in 1959. (Sandi Stockham)

after World War II the Food & Machinery Corporation (FMC), which later became the Food, Machinery & Chemical Corporation, acquired Bolens and the first FMC Bolens compact tractors appeared in 1947.

The four-wheel Bolens Ride-a-Matic garden tractor with a 7 hp Kohler engine was made in America from the mid-1950s and was introduced to the UK market in 1959 by Garden Machinery Ltd of Slough. The Ride-a-Matic, which cost £230, had a belt drive variable-speed Bolens Versa-Matic transmission with two forward speeds and one reverse. The belt was slack when the transmission was in neutral and reverse was engaged with a hand lever used to move the belt on to the reverse drive pulley. The two speed ranges were selected by engaging the high or low gear in the chain drive housing to the rear axle.

2.11. The 6 hp Bolens Husky 800 had a Briggs & Stratton four-stroke engine.

Although the Ride-a-Matic was basically a ride-on rotary mower it could also be used with other attachments such as a plough and a rotary cultivator, making it one of the smallest ride-on tractors in the late 1950s which could be used with a rotary cultivator. Garden Machinery of Slough also imported Bolens Suburban ride-on mowers in the late 1950s.

Introduced in 1961, the Bolens Husky 600 ride-on light tractor with a wide selection of quick-attach implements was the first of a group of Bolens Husky tractors sold in the UK by Mini-Tractors Ltd of Slough. The Husky 600 with a 6 hp Briggs & Stratton engine, a three forward and one reverse gearbox, combined clutch and brake, an implement lift, a belt pulley and direct-drive power take-off cost £215, electric starting adding £30 to the price. Within two years Mini-Tractors was selling the four models of the Bolens Husky tractor including the 6 hp 600 and the

Wisconsin-engined 8 hp Husky 800 and 900 and the 10 hp Husky 1000.

Standard features of the Husky 800, 900 and 1000 included a six forward and two reverse gearbox and the Power-Lock quick-attach system used for power-driven equipment including a rotary mower deck, a gang mower, a rotary cultivator and a snow blower. Implements for the optional Husky hydraulic linkage included a plough, cultivator, disc harrow, rotary brush and a trailer.

The Howard Rotavator Co was appointed sole UK concessionaire for Bolens products in the late 1960s. The tractors were badged Howard Bolens, an arrangement which continued until Howard ceased trading in 1985. Howards was selling Bolens lawn, garden and estate or compact tractors in the early 1970s. There were four Briggs & Stratton-engined lawn tractors, four garden tractors and the Howard

2.12. The Bolens QT16 was known as the quiet one. (Sandi Topham)

Bolens 1476 estate tractor with a 14 hp Wisconsin engine, hydrostatic transmission and hydraulic linkage.

The smallest 853 garden tractor had an 8 hp Briggs & Stratton engine, the 10 hp 1053/4 had a Wisconsin power unit and either a Tecumseh or Wisconsin engine was used for the 12 hp 1237/4. The standard specification included a belt and gear transmission system, six forward and two reverse gears, shoe brakes and a 3,600 rpm power take-off. The Bolens 1257/56 with a 12 hp Tecumseh or Wisconsin engine

had a hydrostatic transmission. Implements for Bolens garden tractors included a rotary mower deck, gang mower, rotary cultivator and a plough.

Howard Rotavators of Bury St Edmunds in Suffolk imported seven 8 to 19 hp Bolens tractors in the mid-1970s. The smallest 8 hp 828 and 829 with Tecumseh engines and the G-8 with a Briggs & Stratton power unit were riding mowers. The Howard Bolens compact tractor range included the Tecumseh-engined 10 hp G-10 with a three forward and one reverse gearbox, the 14 hp six forward and two reverse speed G-14 and the 14 hp H-14 with a hydrostatic transmission.

The QT-16, labelled the quiet one, had a twin-cylinder 16 hp Onan engine and the HT-20 with a 19½ hp Kohler twin-cylinder power unit were both equipped with a hydrostatic transmission, power take-off and hydraulic linkage. A mid-mounted rotary mower deck, rotary cultivator and a plough were among the attachments made for the QT-16 and HT-20. Bolens also distributed the Iseki TX diesel tractors in its chocolate and white livery in the UK from 1978, adding the Iseki TE tractor range in 1984.

Claymore Grass Machinery of Birmingham was appointed the UK concessionaire for Bolens tractors after Howards ceased trading in 1985. Claymore Grass Machinery was a subsidiary of Reekie Engineering Ltd of Arbroath and Bolens tractors were marketed from outlets in Birmingham and Arbroath. The range included 14 and 16 hp Bolens garden tractors, the 12½, 16 and 18 hp Bolens Eurotrac with twin-cylinder Briggs & Stratton engines and the 18 and 20 hp Bolens Estate tractors with Kohler power units.

The Garden and Estate models had cruise control and the Eurotrac range was available with a five-speed manual or a hydrostatic transmission. The 23 hp HT-23 with a Kohler engine and hydrostatic transmission completed the range of Bolens tractors in the UK market in the mid-1980s and early 1990s. All Bolens tractors could be used with a rotary

2.13. The Howard Bolens G-14 had six forward and two reverse gears.
(Sandi Topham)

mower deck and a small trailer and most of them with a rotary cultivator. Claymore Grass Machinery was based at Bidford-on-Avon in the mid-1990s when it sold Bolens Suburban lawn tractors and pedestrian-controlled rotary mowers.

The Bolens product division of Milwaukee Equipment, a subsidiary of FMC, was bought by The Garden Way Corporation (Troy-Bilt) of Port Washington in the 1980s. Garden Way filed for bankruptcy in 2001 and its assets were bought by MTD (Moll Tool & Die Products) of Cleveland, Ohio.

Interestingly, when a large Ford and Bolens tractor dealer network in America lost the Ford tractor franchise in the early 1960s it agreed to become main dealers for David Brown tractors on condition that the Meltham-built tractors were painted in the Bolens chocolate and white colour scheme. While on a visit to America a group of UK David Brown dealers saw the new colour scheme and requested the same paintwork for the British market. The first chocolate and white David Brown tractors were sold to British farmers in October 1965.

BRISTOL

Roadless Traction Ltd of Hounslow conceived the idea of a lightweight crawler tractor and built some prototype machines with Douglas flat-twin engines and rubber-jointed tracks in the early 1930s. Douglas Motors meant to build the first crawlers in its factory at Kingswood near Bristol, the location giving the tractor its name. However, Douglas Motors went into receivership before any tractors were built so Bristol Tractors Ltd was established in 1933 and the first Bristol crawler tractors were made at Willesden in London.

The first batch of bull-

nosed Bristol crawlers with 7 in wide Roadless Traction rubber-jointed tracks had 1,350 cc air-cooled British Anzani V-twin engines and a three forward and one reverse gearbox. They were steered with a single tiller lever which controlled the independent differential brakes on the front sprocket drive shafts. A twist grip throttle on the steering tiller controlled engine speed.

Publicity material described the Bristol crawler as a modern farm tractor which could be worked day and night at less cost than the upkeep of two horses. The Bristol was 6 ft 6 in long, less than 3 ft wide, weighed one ton, used a gallon of petrol an hour and cost £155 when delivered to the nearest railway station.

There were problems with the British Anzani power unit so from 1934 buyers had the choice of a water-cooled Jowett flat-twin petrol engine or a 10 hp Coventry Victor diesel engine. Depending on customer requirement the tractor was supplied with an overall width of between 35½ and 60 in.

Within a year the Bristol Tractor Co was in financial difficulty and the Jowett Car Company bought the business in 1935. Tractor production was transferred

2.14. The Bristol 10 had Roadless Traction rubber-jointed tracks. (Roger Smith)

2.15 The Bristol 20 crawler was introduced in 1948.

to the Jowett car factory at Idle near Bradford in Yorkshire when the Bristol crawler, with a Jowett horizontally opposed twin-cylinder petrol engine, cost £195.

A four-cylinder Jowett engine with two pairs of horizontally opposed cylinders was introduced in 1937 and the original bull-nosed bonnet was replaced with a flat-fronted radiator grille. An optional twin-cylinder horizontally opposed Victor Cub diesel was introduced a year or so later.

An industrial version of the 10 hp Austin four-cylinder petrol or tvo engine was used for the Bristol crawler from 1942. The petrol/tvo engine suffered from unburnt tvo diluting the engine oil but this problem was overcome in 1944 by using separate petrol and tvo carburettors. The seat, exhaust pipe and air cleaner were lowered to reduce the overall height and track shields were added for the orchard version of the Bristol 10 introduced in the same year.

The last Bristol 10 tractors with a Union flag on the bonnet were made in 1947, when press advertisements carried the slogan 'Bristol crawlers are British made in British factories for British people'.

Jowett sold Bristol Tractors to the Austin car distributor H A Saunders of Finchley which transferred production of the Bristol crawler to Earby near Colne in Lancashire. The Bristol 20, introduced in 1948, was three inches wider than the previous model and had a modified 22 hp Austin 16 overhead-valve car engine, a single-plate clutch, a three forward and one reverse gearbox and spur gear final drive.

The petrol and tvo versions of the Bristol 20 both cost £480 and electric starting added £22 10s 0d to the price. The tvo engine had two Zenith carburettors to prevent the fuel diluting the engine oil. The Roadless rubber-jointed tracks were retained, steering was by multi-plate clutches at the rear-end controlled by two hand levers and independent foot brakes were provided to assist with sharp headland turns. The power take-off shaft was standard but an extra £47 10s 0d was added to the bill for a bolt-on rear hydraulic unit and three-point linkage with the hydraulic pump driven from the gearbox lay shaft.

An Austin A70 industrial engine replaced the Austin 16 engine during the latter part of the Bristol 20's production run which came to an end with the launch of the Bristol 22 in 1952. It was similar to the Bristol 20 with an Austin A70 petrol/tvo engine and improved track gear. The track setting was adjustable from 30 to 44 in on tractors with 7 in wide plates and from 33 to 47 in with 10 in wide track plates. An alternative 23 hp Perkins P3 diesel engine was made available in 1953 when the basic Bristol 22 cost £742.

More powerful Bristol crawlers appeared in the mid-1950s but most were too powerful for the smallholder and market gardener. The Bristol 25 with the option of a 22 hp petrol/tvo engine or a Perkins P3 was introduced in 1956. The new model had a three forward and one reverse gearbox, hand-lever operated dry multi-plate steering clutches and independent foot brakes. The specification included a six-spline power take-off, belt pulley and category I three-point linkage suitable for use with Bristol, David Brown and Ferguson implements.

Bristol crawlers in the 1960s included the 32 hp PD (Power Diesel) series with a Perkins engine which cost £1,100 in 1967, and the PD44 and PD48 which were industrial tractors used mainly with a dozer blade or front-loading shovel.

The 40 hp Bristol Taurus with a Perkins diesel engine and a six forward and four reverse gearbox cost £1,800 in 1967. The Thomas Ward Group, which owned Marshalls of Gainsborough, bought the Bristol tractor business in 1970 when the industrial version of the Bristol Taurus became the Track Marshall 1100.

BYRON

In 1946 Byron Farm Machinery at Walthamstow placed an advertisement in the farming press advising growers and smallholders that it would be ready to accept orders for delivery of the new three-wheel Byron rowcrop tractor the following year. The Byron Mk I, like other tractors of the day, had an industrial version of the popular four-cylinder water-cooled Ford 10 petrol engine with 6 volt coil ignition and electric starting. The tractor had a single-plate clutch, a three forward and one reverse gearbox with a top speed of 8 mph and independent rear wheel brakes with a parking latch.

Sales literature pointed out that the combination of a single front wheel and independent rear wheel brakes not only gave excellent manoeuvrability but also helped to keep the price at the competitive level of £260 on steel wheels and £280 on

pneumatic tyres. A hand-operated mid-mounted toolbar was standard and it was claimed that the Byron had enough power to pull a two-furrow plough or a 5 ft cultivator in normal working conditions.

Introduced in 1949, the modified Byron Mk II tractor was recognisable by its slimmed down mudguards designed to improve visibility for the driver, especially when working in rowcrops. The standard petrol-engined version of the Mk II tractor with improved brakes and steering cost £247 10s 0d and the rowcrop version which included a 7 ft wide mid-mounted toolbar was £292. An optional tvo conversion kit cost £56, a pair of 4½ in wide steel spud wheels for working in narrow rows was £25 and a set of lights cost £10 15s 0d. The rear wheel track was adjustable from 56 to 73 in and a side-mounted belt pulley was available as an optional extra.

Implements for the Byron tractor included mid- and rear-mounted toolbars raised and lowered with a spring-assisted hand lever, a plough and a set of disc harrows. The tvo-engined Byron cost £348 when it

2.16. The Byron tractor on steel wheels cost £260 in 1947.

was discontinued in 1954. When compared with the price of £335 for the Ferguson TED 20 it was doubtful if the Byron offered real value for money.

CAPLIN ENGINEERING

Introduced in the mid-1960s the Capco Fieldrider 301 was made by Caplin Engineering of Ipswich for about ten years. The Fieldrider had a 147 cc Norton Villiers F15 four-stroke petrol engine with a bright red fibreglass bonnet and electric starting was an optional extra. The transmission consisted of a complex arrangement of vee-belts, four forward gears and one in reverse, differential and a transverse

rear axle. The vee-belt pulleys ran on sealed-for-life bearings. An over-centre jockey pulley served as the clutch, and a quadrant-and-pinion steering box gave the tractor a 52 in turning radius.

2.18. The Fieldrider had a complex vee-belt transmission system.

An optional Tarpen flexible power take-off drive could be used with various tools including a hand-held rotary cultivator, hedge trimmer and a pruning saw. The Fieldrider also had hitching points for seed and fertiliser spreaders, a gang mower, roller, dump cart, rake, sprayer and other attachments. The 5 hp Fieldrider light tractor with an overall width of 32 in. A full range of mounted and trailed equipment was listed in 1974 by the Fieldrider Division of Caplin Engineering which was then part of the Manganese Bronze group.

CARTERSON

Horace Carter and his son designed and built the first Carterson tractor with a single-front wheel in 1949 but the tricycle layout was soon changed to the more conventional four-wheel arrangement. Front-, mid- and rear-mounted hand-lift toolbars were made for the three-wheeler but the four-wheel version was mainly used as a towing tractor. The Carterson had a

CAPCO

Fieldrider
The new low-cost, high-output mini tractor. Goes anywhere-Does anything-For anybody

2.17. The Capco Fieldrider 301 cost £285 in the mid-1960s.

2.19. The Carterson tractor was built at Northwich in Cheshire. (Brian Carter)

kickstart single-cylinder Norton side-valve engine which developed 8½ bhp at 2,600 rpm, rising to 16 bhp at 4,700 rpm at full throttle. The three forward and one reverse gearbox provided top speeds of 2½, 5 and 10 mph.

It was steered by a system of pulleys and wire rope and an unusual clutch control mechanism returned the engine to tick-over when the pedal was depressed. Independent rear wheel drum brakes gave the Carterson a 4 ft 6 in turning radius and the wheel track was adjustable with spacers on the rear axle in 3 in steps from 36 to 60 in.

The tractor, which was suitable for rowcrop work, could also be used with a conventional or reversible plough and a drawbar was provided for towing a trailer. The Carterson cost £249 in 1949 when press publicity explained that it was anticipated that growers would be able to buy the tractor and a full set of attachments for well under £300.

Fifteen Carterson tractors were built at Northwich in Cheshire in 1949 and 1950. Five further sets of tractor parts were made but they were sold off unassembled and their eventual fate is not known.

COLEBY SHIRE

Frederick Coleby built the Coleby Shire four-wheel tractor at Swanley in Kent in the mid- to late 1950s. The first Shire tractors had a twin-cylinder air-cooled 10 hp JAP engine, a three forward and one reverse gearbox with a top speed of 3 mph, a pedal-operated transmission brake and independently driven rear wheels. The Shire had a power take-off, belt pulley and an adjustable wheel track but as it did not have a differential it was necessary to disengage the drive to one wheel before turning at the headland. An unconventional hydraulic linkage unit had a piston pump to supply oil to a vertical lift ram mounted on the drawbar and connected to the top link.

The New Light Weight, Hydraulic Operated "SIX" COLEBY

B. G. PLANT (Sales Agency) Ltd. DISTRIBUTORS

2.20. The Coleby Shire.

Various mid- and rear-mounted implements, including a plough and a cutter bar mower, were made for the tractor. The Shire Six, a later version of the Coleby tractor, appeared in 1958. It had a 420 cc BSA Model G air-cooled four-stroke petrol engine, rated at 5¾ hp, a single-plate clutch and a three-speed power take-off.

CRAWLEY

Crawley Metal Productions made about 200 Crawley 75 tractors in the late 1950s at its factories in East Preston near Littlehampton and Winschoten in Holland. The Crawley 75 was a British development of an American design that had proved itself over a number of years. The basic tractor was made in Holland and Crawley Metal Productions modified it to suit British requirements when the Crawley 75 was introduced to the UK market in 1958. Spare parts and accessories for the tractor were made at Little

Preston and dispatched to customers from the nearby Angmering railway station.

Buyers were given the choice of an air-cooled 8 bhp JAP model 5B petrol engine, a 5BV vaporising oil engine or a a 7 bhp Petter PC1 Mk II air-cooled diesel. Most Crawley 75 tractors were sold with a Petter engine, which used about 1½ gallons of diesel in a ten-hour day, compared with the daily consumption of two gallons by the petrol and tvo engines.

The Crawley 75 had an automatic centrifugal clutch, three forward gears and one reverse, a worm-and-wheel differential assembly and independent rear wheel brakes. The specification also included draft control hydraulics with a belt-driven oil pump, a central power take-off shaft and a universal drawbar. The wheel track was adjustable in four steps from 32 to 48 in by reversing one or both wheels on their hubs. Optional equipment included a foot throttle, a rear power take-off adapter and front wheel weights.

2.21. The Crawley 75 was designed in America.

The Crawley 75 cost £280 and the diesel model £298 when advertised in 1958 as the first tractor of its size sturdy enough to stand up to heavy-duty work. Publicity material suggested that every farm should have a Crawley as it could do every job around including ploughing, drilling, spraying, fertiliser spreading, gang mowing and hay making with a mid-mounted cutter bar mower. Buyers were also tempted with an offer of two free services for their Crawley tractor during the first eighteen months of its life.

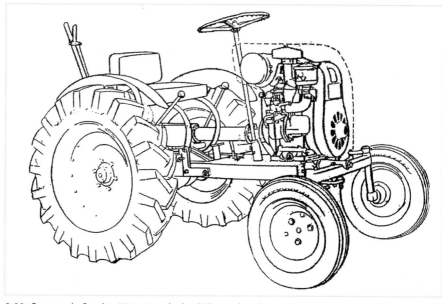

2.22. Some early Crawley 75 tractors had a JAP petrol engine.

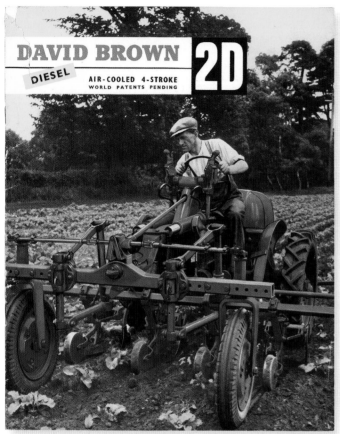

2.23. The twin-cylinder Air-Light implement lift system was an unusual feature of the David Brown 2D

The basic specification of the 2D included a single-plate dry clutch and a four forward and one reverse gearbox mounted above the rear axle. Shoe brakes with independent pedals were carried on the half shafts and the wheel track was adjustable from 40 to 68 in.

The driving seat in front of the engine afforded a clear view of the work being done by the underslung toolbar which was raised and lowered by the David Brown Air-Light lift system. This unusual design employed an air compressor, driven by the tractor engine, to supply compressed air at 120 to 130 psi to a pair of lift-cylinders linked to the toolbar. The tractor's 4 in diameter tubular chassis served as the air reservoir, and a tapping-off point for inflating tyres was an added bonus. The 8 or 10 ft mid-mounted toolbar was attached to the 2D by fitting both of its depth wheels at one end of the bar and then wheeling it into position under the chassis. An optional rear-mounted toolbar with its own airlift ram was added in 1958.

A forward-facing power take-off shaft from the front of the gearbox with its speed related to the tractor's ground speed was standard but a live rear power shaft with epicyclic gears controlled by a brake band operated through the Air-Light system was an optional extra. Other optional equipment included a rear belt pulley and electric lighting.

DAVID BROWN 2D

In 1955 David Brown Tractors launched the rear-engined 2D self-propelled tool carrier at the Royal Smithfield Show. The first tractors came off the Meltham Mills production line in 1956, and in 1957 a narrow vineyard model was added. The four-stroke air-cooled twin-cylinder David Brown diesel engine with a third balancing piston and cylinder in the sump developed 12 bhp at 1,500 rpm but within a year sales literature quoted engine power as 14 bhp at a rated engine speed of 1,800 rpm.

2.24. The lift rams were used independently when reversible ploughing with the 2D. (Roger Smith)

Numerous implements, some made by David Brown and others approved for use with the 2D, included hoe blades, cultivator tines and ridging bodies for the underslung and rear toolbars, a down-the-row thinner, mid-mounted mower and a reversible plough. Although claimed to be a maid of all work the tractor was more suited to rowcrop work. Just over 2,000 2D tractors, including about 400 vineyard models, had been made when production came to an end in 1961.

GARNER

The four-wheel Garner light tractor, developed from the earlier two-wheel model, cost £197 15s 0d on steel wheels when it was introduced by Garner Mobile Equipment in 1949, pneumatic-tyred wheels adding an extra £5 to the price. The new tractor had features in common with the two-wheel garden tractor. These included the 5/6 hp JAP Model 5 air-cooled engine, centrifugal clutch, three forward and one reverse gearbox, spiral bevel gear and differential, roller chain final drive and expanding shoe brakes.

The throttle was conveniently placed on the steering column in front of the driver who sat with his back to the engine. However, the gear lever was in the same position as it was on the pedestrian-controlled tractor and the driver had to reach behind his back to change gear. The three forward gears gave a speed range of ½ to 10 mph and the wheel track was adjustable from 25 to 42 in. The Garner had a power take-off, belt pulley and drawbar, while the mid-mounted toolbar and rear-mounted implements were raised and lowered with a hand lever.

The original Garner was considered to be rather under-powered so from 1950 a slightly longer and wider tractor with a 7 hp JAP Model 6 petrol or tvo engine which cost £208 15s 0d on steel wheels was built alongside the 5/6 hp tractor. Implements for the Garner light tractor included a mid-mounted

2.25. The Garner light tractor. (Roger Smith)

one-way plough attached to the tractor in the same way as the mid-mounted toolbar and the usual range of cultivation equipment, seeder units, potato and sugar beet lifters.

A haulage version of the Garner was exhibited at the 1953 Smithfield Show and some, with a 7 or 10 hp JAP 55 twin petrol engine and a top speed of 8 mph, were used by the Docks and Inland Waterways Executive. Production of the four-wheel Garner tractor ceased in 1955.

GROWMOBILE

Evenproducts of Evesham, which was well known for its irrigation equipment, introduced the Growmobile mechanised tool carrier in the mid-1980s. There were two models of the Growmobile with rear-mounted air-cooled Honda petrol engines, hydrostatic motors in the rear wheels and top speeds of 4 mph in forward and in reverse. The smaller 8 hp Mk I-R Growmobile with a 26 in ground clearance cost £4,675 in 1987 while the 11 hp Mk II-R with a 33 in ground clearance was £5,225.

The separate operating levers used to control the speed and rotational direction of the hydrostatic motors made the steering wheel redundant. Moving both levers forward at the same rate kept the machine on a straight path and moving them

2.26. An electric lift mechanism was used to raise and lower the Growmobile's underslung toolbar.

The three-wheel Gunsmith, designed by Harold Smith, had the same 6 hp air-cooled engine and two-speed transmission as the BMB Plow-Mate. The three-wheeler, which took its name from Norman Gunn and Harold Smith who were the directors of Farm Facilities at Maidenhead and Twickenham, cost £178 when it was introduced in 1948.

The first Gunsmith tractors assembled at Maidenhead were built with engines, gearboxes and rear axle units bought from Brockhouse Engineering. Some Mk I Gunsmith tractors had a Briggs & Stratton ZZ four-stroke engine but most were equipped with a single-cylinder 6/7 hp JAP air-cooled engine.

backwards put the Growmobile into reverse. A slight change of direction while in work was achieved by moving one lever slightly further forwards or backwards and a more pronounced movement of the levers facilitated turning at the headland.

The underslung toolbar used for cultivation, hoeing, drilling and other rowcrop work was raised and lowered by an electric lift mechanism powered by a 12 volt battery charged by a heavy-duty alternator. Separate electric motors were used to drive a fertiliser spreader, a granule applicator, a sprayer or a weed wiper mounted at the front or rear of the machine. This arrangement allowed the operator to raise or lower the toolbar and control a fertiliser spreader or other attachment at the same time.

Power was transmitted by belt from a pulley on the engine crankshaft to a two forward and one reverse gearbox. Stepped flat-belt pulleys on the engine output and gearbox input shafts provided a high-low ratio arrangement to give a top speed of 3 mph in low ratio and 5 mph when using the larger diameter side of the engine crankshaft pulley. Drive was engaged with a pedal-operated jockey pulley clutch, which tensioned the driving belt. The manufacturers

GUNSMITH

Sales literature suggested that the Gunsmith tractor, manufactured by Farm Facilities, would do its work as well as any big tractor, often more precisely and always more cheaply. It was also suggested that youngsters would find it so easy to operate they would delight in doing men's work.

2.27. The first Mk I Gunsmith light tractors were made in 1948.

2.28. The Mk II Gunsmith had a steering wheel.

pointed out that by merely replacing the belt this type of clutch could be overhauled in a matter of seconds with no mechanical knowledge required.

The Gunsmith front wheel was handlebar-steered and had a turning radius of 5 ft when assisted by pedal-operated differential brakes. The driving seat positioned in front of the engine gave an uninterrupted view of the mid-mounted toolbar which, like the plough and rear-mounted toolbar, was raised and lowered with a hand lever.

The Gunsmith had a maximum drawbar pull of about 800 lb, the power take-off could be used to drive a tractor-mounted generator or pump and the optional belt pulley provided a maximum belt speed of 2,200 ft/min. In addition to the usual range of cultivation implements the equipment list for the Gunsmith included a trailer, sprayer, air compressor and a pneumatic hedge trimmer, saw bench and the Farfac-Scotmec hammer mill.

The Mk II Gunsmith cost £197 10s 0d when it made its debut at the 1951 Royal Smithfield Show. The most obvious change was its steering wheel with a 3:1 ratio sun-and-ring gear instead of handlebars. Optional four-groove vee-belt pulleys for the engine and gearbox provided additional forward and reverse gears and a top speed of 8 mph. The Mk II

was slightly wider and longer and stood 6 in higher than the Mk I tractor. It had an improved floating drawbar and the wheel track was adjustable from 30 to 58 in with axle extensions.

New FarFac accessories also exhibited at Earls Court in 1951 included a tipper skip, double tandem disc harrow, side-mounted cutter bar and the Gray Smallholder belt-driven hammer mill mounted above the tractor seat.

Apart from an enclosed sliding mesh gearbox with four forward and two reverse gears there were no other obvious changes in the specification for the Mk III Gunsmith listed in a 1958 farm equipment directory. There was also a four-wheel version of the Gunsmith, but this was probably a modification by Garden Machinery Ltd which had recently purchased the remaining stock of Gunsmith parts from Farm Facilities.

About three hundred Gunsmith tractors were made during its six year production run and many of them were exported, mainly to Australia, Canada and South America. Satisfied users wrote testimonial letters in praise of the Gunsmith. One owner described it as a wonderful machine that would do five hours' work in one hour while he sat in an armchair – his description of the Gunsmith's seat.

A Wigan smallholder reported that he had only used three gallons of petrol to rake up and cart eight acres of oats. Another wrote to say that a boy of sixteen was able to do everything required on his holding with the Gunsmith tractor.

2.29. The Howard Mini-Gapper.

HOWARD

The Howard name appeared from time to time on ride-on horticultural machines. A Howard Clifford six-wheel drive tractor and log trailer was demonstrated to the Forestry Commission in 1960. The articulated tractor unit had a 12 hp petrol or diesel engine, a four forward and two reverse gearbox with a top speed of 4½ mph, and the wheels on the log trailer were driven from the tractor's power take-off shaft.

Mechanical thinning of the sugar beet crop was in fashion when the Howard Rotavator Co introduced the farmer-designed self-propelled Howard Mini-Gapper in the early 1960s. The Mini-Gapper was used for cross blocking, a task previously done with a horse

hoe, which entailed hoeing across the rows at a right angle to leave small clumps of plants to be singled by hand.

The rear-engined Mini-Gapper had a single-cylinder 3.2 hp Kohler power unit with a recoil starter, a hand clutch to tension the vee-belt drive and a diff-lock. Four forward and four reverse speeds were provided by a two-speed vee-belt drive and a high-low ratio gearbox. The Mini-Gapper had a steering wheel and a transmission brake. The hoe bar was raised and lowered with a hand lever. Foot pressure could be applied to the hoe bar to improve penetration when working in hard ground.

The Howard Rotavator Co was also the sole concessionaire for the American Bolens garden tractors (page 98) from the late 1960s until it ceased trading in 1985.

INTERNATIONAL

The International Harvester Co of Great Britain imported the American-built International 7 hp Cub Cadet in the early 1960s. It had a single-cylinder four-stroke petrol engine and combined a single-plate clutch and a brake operated by a single two-stage pedal. The first stage disengaged the clutch

2.30. A late 1960s International Cub Cadet.

and the second applied the brake. The transmission consisted of a vee-belt drive to a three forward and one reverse gearbox and direct drive from the differential to the rear axle. The Cub Cadet had car-type worm-and-gear steering, front and rear power take-off shafts and a hand-lift lever with six positions for height adjustment was used for the front and rear toolbars and mid-mounted mower.

The International Cub Cadet 70 and Cub Cadet 100 were current in the mid-1960s. The Cub Cadet 70 had the same 7 hp Kohler engine as the previous model and a 10 hp Kohler was used for the Cub Cadet 100. A recoil starter was standard although electric starting was available at extra cost. The basic specification for both models was very similar to the 7 hp Cub Cadet.

Optional equipment included a front power take-off shaft, a slow creep-speed gearbox and a hydraulic lift attachment that simplified the use of the hand lever-operated mower and toolbars. A full range of implements was made for the Cub Cadet including a plough, seeder unit, sprayer, rotary sweeper, front-mounted saw bench and turf care equipment.

International Harvester was importing three models of Cub Cadet in the mid-1970s. The smallest 10 hp tractor had a three-speed gearbox. There was a choice of a three-speed manual gearbox or hydrostatic transmission for the 12 hp model, and hydrostatic transmission was standard on the 14 hp Cub Cadet. Cub Cadet tractors were imported in the early 1980s by Marshall Concessionaires at Brackley in Northamptonshire which also marketed Textron Jacobsen mowers.

ISEKI

The Iseki Agricultural Manufacturing Co was established in Japan in 1962 but Iseki four-wheel tractors did not arrive in Great Britain until 1976. Following a visit to Japan, the Suffolk John Deere tractor dealer Len Tuckwell started a company called

2.31. L Toshi Ltd imported the first four-wheel drive Iseki TX 1300 compact tractors in 1976

2.32. Iseki UK imported the 16 hp Iseki TX2140 in the mid-1980s.

2.33. Jacobsen, a member of the Textron group, became the UK distributors of Iseki tractors in 1997.

The same models were current in 1979 when a Lely-Iseki partnership based at St Neots in Cambridgeshire replaced L Toshi as the UK distributor of Iseki compact tractors. The Lely-Iseki price list for 1982 included the two-wheel 28 hp TS 2810, the two- and four-wheel drive TX 1300 and TX 1500F, together with the four-wheel drive 18 hp TS 1910F and 31 hp TS 3110 compact tractors.

There was no regularity in the number of gears provided; the TS 1910F had twelve forward and six reverse gears but there were only eight forward and two reverse ratios in the TS 3110 gearbox. The two- and four-wheel drive 14½ hp TX 2140 and 16½ hp TX 2160 tractors were a later addition to the Lely-Iseki range. The standard specification included three-cylinder water-cooled engines, a six forward and two reverse gearbox, three-speed rear and single-speed front power take-off shafts, hydraulic linkage and diff-lock.

The Lely-Iseki agreement ended in 1986 when Iseki UK Ltd opened new premises at Little Paxton near Huntingdon before moving to Bourn in Cambridgeshire. The Iseki compact range at the time included the TX 2140 and 2160 and five medium-sized TE tractors from the 24 hp TE 3210 to the 48 hp TE 4451. The three-cylinder TX 2160 was also available with two-speed range hydrostatic transmission and two-speed power take-off, a backhoe and a front loader bucket.

It was all change again in 1993 when, following a marketing agreement between Massey Ferguson and Iseki, a range of MF tractors in Iseki livery was sold in Japan with an Iseki badge. Some models of Iseki compact tractor were also sold in Great Britain with the MF logo and red paintwork.

The Massey Ferguson agreement continued until 1997 when the Textron Group at Jacobsen House in Kettering took over the sales of the four-wheel drive Iseki TM and TF compact tractors in the UK. There was a choice of a six forward and two reverse manual gearbox or a two-range hydrostatic transmission for the TM 215 and TM 217 with 15 and 17 hp water-cooled diesel engines. The 20 hp TF 321, 25 hp TF

L Toshi Ltd in Ipswich in order to import Iseki compact tractors, including the 15 hp TS 1500, the 28 hp Iseki TS 2810 and the 13 hp four-wheel drive TX 1300F. Within a couple of years the Toshi range had been extended to include the two-wheel drive TS 1300, TS 1910, TS 2110 and TS 3510. The TX 1300 and TX 1500 tractors and a two-wheel drive version of the TX 1300F had twin-cylinder Mitsubishi diesel power units and a six forward and two reverse gearbox.

The specification included a three-speed power take-off, diff-lock and, in common with all Iseki tractors of the day, a safety start device was built into the clutch pedal linkage. A three-cylinder four-stroke water-cooled Isuzu diesel engine was used for the 28 hp TS 2810 and 35 hp TS 3510. A nine forward and three reverse gearbox was used on the TS 2810 and the TS 3510 had eight forward gears and two in reverse.

325 and 30 hp TF 330 had three-cylinder indirect injection diesel engines with the option of a manual or hydrostatic gearbox.

Following the acquisition of Ransomes by Textron Inc of America in 1998 the Jacobsen Iseki range of compact tractors was distributed from Ipswich.

JOHN DEERE

2.34. The John Deere 110 was the first compact model sold in the UK.

The John Deere 110 garden tractor, apparently made because the company wanted a machine to mow the grass outside its global headquarters in Moline, was built at John Deere's Horicon Works in Wisconsin in 1963. The 7 hp John Deere 110 had a Kohler single-cylinder four-stroke petrol engine, a four-speed forward and reverse gearbox with a belt and chain to the rear wheels. It had a top speed of 6½ mph and drive was engaged with an over-centre pulley arrangement used to tension the belt from the engine to the gearbox.

Lundell at Eastbridge in Kent sold John Deere farm equipment in the UK until 1966 when the American manufacturer established an agricultural machinery depot at Langar near Nottingham. Garden tractors and lawn mowers were not included and Stanhay, based at Ashford in Kent, which was appointed concessionaire for John Deere ground care equipment, introduced the 7 hp John Deere 110 garden tractor in 1966. Stanhay introduced the 14 hp John Deere 140 with hydrostatic transmission in the early 1970s but its involvement with John Deere compact tractors ended in 1972 when Ground Control Ltd took over the marketing of John Deere ground care equipment.

Ground Control, based at Liphook in Hampshire, included seven models of John Deere garden and estate tractor in its price list for 1976. The 8 hp 100 series with a Briggs & Stratton engine, a three forward and one reverse gearbox and a 34 in rotary mower deck was priced at £882. The John Deere 200 series with 8, 10, 12 and 14 hp Kohler engines had four forward gears and one reverse, a variable speed drive and either a manual or hydraulic lift.

Air-cooled Kohler petrol engines were also used for the 16 hp 300 Series tractor and 19.9 hp twin-cylinder 400 Series model. Both had hydrostatic transmission

2.35. The 17 hp John Deere 185 Hydro was one of six 100 Series compact tractors in the late 1980s.

with single lever control and hydraulic linkage, additional features of the 400 series tractor included a high/low axle ratio and power steering. A rotary mower deck and a rotary tiller were available for all mid-1970s John Deere compacts and a cutter bar mower, a flail mower, a front dozer blade and a single-furrow plough were among the available attachments for the 300 and 400 series tractors.

When the Ground Control business failed in the late 1970s remaining stocks of John Deere garden tractors and equipment were taken to Langar and sold by Greenlay (Grass Machinery) of Northumberland. John Deere established a Groundcare Division at Langar in 1986 and Greenlay Ltd became the John Deere ground care equipment dealer.

KENDALL

After a period of field testing, plans were put in place to manufacture up to fifty three-wheel Kendall-Beaumont tractors by the end of 1945. The tractor was developed by Denis Kendall MP, and Beaumont refers to the cars and motor cycles with 6 hp three-cylinder radial engines designed by Horace Beaumont in the 1920s and 1930s.

The prototype Kendall tractor had the same radial engine but it lacked power so a supercharged version rated at 7 hp was used for the first production models built by Grantham Productions in Lincolnshire. Even then the radial engine was not a great success, so after considering the use of a Douglas or a Petter flat twin engine, it was decided to fit an 8 hp air-cooled

2.36. The first Kendall tractors had a three-cylinder radial engine.

twin-cylinder Douglas petrol engine.

The Kendall three-wheel tractor, which cost £100 in 1946, was suitable for light field work at speeds of up to 6 mph and for road haulage at 20 mph. Within a few months Grantham Productions went into liquidation and Newman Industries at Yate near Bristol bought the Kendall assets and the Grantham factory. After refurbishing the premises the new owners introduced the Newman three-wheel tractor in 1948.

KENT PONY

Outboards Ltd of Whitstable in Kent made the Kent Pony tractor in the early 1950s. Like many other light tractors of the time it had a Ford 10 industrial petrol engine, although a tvo conversion kit was available at extra cost. The tractor had an extra-large capacity radiator to keep the engine cool when it was used for long periods of stationary work. Power was transmitted to the rear wheels through a single-plate clutch, a three forward and one reverse Ford gearbox with a top speed of 15 mph and worm-and-wheel drive to the rear axle.

The standard specification included independent rear wheel brakes and power take-off while the drawbar was attached under the rear axle housing. Narrow and wide track settings were obtained by reversing the wheels on their hubs.

KUBOTA

The Marubeni Corporation made its first pedestrian-controlled Kubota garden cultivators and introduced four-wheel Kubota compact tractors to the Japanese market in 1960. By

2.37. The Kent Pony was made at Whitstable.

2.38. The Kubota B6000 had a 12½ hp twin-cylinder water-cooled diesel engine.

the end of the decade a considerable number of diesel-engined Kubota compact tractors had been sold in America.

Kubota tractors appeared in the UK in 1975 when the first B6000, B6000E, L175 and L223 compact models were sold from premises at Whitley Bridge in North Yorkshire. A range of matched equipment was already available for the B6000 and similar attachments for the L225 and L175 were at the development stage.

The four-wheel drive Kubota B6000, which cost £1,535, had a 12½ hp twin-cylinder water-cooled diesel engine, six forward and two reverse gears, front and rear power take-off shafts and independent rear wheel brakes. Optional equipment included category O and I hydraulic linkage, front and rear wheel weights and a rear toolbar.

The two-wheel drive B6000E, which cost £1,387, was in all other respects virtually identical to the four-wheel drive B6000. The two-wheel drive Kubota L175 with a 17 hp twin-cylinder water-cooled diesel engine had an eight forward and two reverse gearbox, independent rear wheel brakes, a single-speed front power take-off and a two-speed power shaft at the back. A diff-lock, category I hydraulic linkage and a safety cab were optional extras for the agricultural version but the diff-lock was not supplied on tractors with turf tyres.

An advertisement in 1975 described the L175 as a tough little tractor that did not act its size. Apart from a 24 hp three-cylinder water-cooled diesel engine the two-wheel drive L225 was very similar to the L175. Complete with a safety cab it cost a little over £3,000. A four-wheel drive version of the L225, designated the L245, appeared in 1977 and the three-cylinder 16

hp four-wheel drive B7100 with six forward and two reverse gears was added to the Kubota compact tractor range in 1978.

Kubota was still at Whitley Bridge when the two- and four-wheel drive 34 hp L345 with a similar specification to the L225 and the B6100 were launched in the UK at the 1980 Royal Smithfield Show. Improvements compared with the L245 included power steering, wet disc brakes and a quiet cab with a detachable roof panel.

Having outgrown its North Yorkshire premises Kubota moved to Thame in Oxfordshire in 1982 when the new two- and four-wheel drive L275 and B8200 tractors were added to the existing range, which included the L245, the L345 and the B7100. The 27½ hp three-cylinder L275 had eight forward and seven reverse gears and category I three-point linkage. The B8200 specification included a three-

2.39. Attachments for the Kubota L225 included a front-end loader, mid-mounted mower and numerous rear-mounted implements.

2.40. A central power take-off shaft was used to drive the Kubota B4200 mid-mounted mower.

2.41. The 15 hp Kubota B6200 had six forward and two reverse gears, a two-speed rear power take-off and a single-speed front power shaft.

cylinder 19 hp water-cooled diesel engine, a nine forward and three reverse gearbox, front and rear power take-off shafts and there was provision to attach a mid-mounted toolbar.

The improved L345 Mk II appeared in 1983 and further introductions in 1984 gave Kubota nine compact models from the 12 hp two-wheel drive B5100 to the two- and four-wheel drive 45 hp L4150. An HST hydrostatic transmission was a new option for the B7100 and B8200. A five-cylinder engine powered the L4150 with an eight forward and eight reverse gearbox, category I and II hydraulic linkage, wet disc brakes and an optional cab. Many of these tractors, including the B4200, B6200 and B7100, remained in production through the 1980s and some were still being made in 1993.

LANDMASTER-GILSON

The Landmaster Division of Boscombe Engineering at Poole was marketing the Landmaster-Gilson ride-on mowers and mini-tractors in the mid-1970s. There was a choice of a 6½ or 8½ hp Briggs & Stratton engine with electric starting for the three forward and one reverse gear tractor with a 25 or 30 in cut rotary mower. The S16 M mini-tractor with a variable four-speed transaxle and the S16 H with a single-lever hydrostatic transmission had 16 hp Briggs & Stratton engines.

The S16 M was equipped with a manual implement lift. The

S16 H had a hydraulic linkage for various implements including a plough, a cultivator, a harrow fertiliser spreader and a dozer blade. Side and rear ejection mid-mounted rotary mowers were also made for Landmaster-Gilson mini-tractors.

LANZ ALLDOG

2.42. Levertons at Spalding sold the Lanz Alldog tool carrier in the mid-1950s. *(Stuart Gibbard)*

The 12 hp petrol-engined A1205 Alldog tool carrier introduced by Heinrich Lanz at Mannheim in Germany in 1951 was superseded the following year by the more powerful Lanz Alldog A1305 with a 13 hp air-cooled single-cylinder two-stroke diesel engine. H Leverton & Co of Spalding introduced the rear-engined four-wheel Alldog with an open tubular chassis to British growers in 1954.

The Alldog specification included a six-speed transmission with a top speed of 12 mph, hydraulic power lifts for mid- and rear-mounted implements, and front and rear power take-off shafts with a belt pulley attachment. The wheel track setting was adjustable from 4 ft 2 to 6 ft 8 in.

Land wheel-driven front and rear power take-off shafts for use with a seed drill or fertiliser spreader were an unusual feature of the Lanz Alldog. The front power shaft was driven by a pair of bevel gears on the right-hand front wheel hub and drive to the rear shaft was taken from final drive gear on the right-hand side of the rear axle.

The wide range of implements made for the Alldog were attached to a series of holes on both sides of the chassis. There was also a purpose-built Allman sprayer with a 150 gallon tank, power take-off driven pump and a 31 ft 6 in spray bar.

The fifth and final model of the Alldog appeared in 1956. Designated the A1806, it had an 18 hp MWM water-cooled diesel engine. John Deere acquired the Lanz business later in 1956 and production of the Alldog continued until 1959 in the John Deere green and yellow livery instead of the earlier Lanz blue paintwork.

LISTER GOLD STAR

The Goldstar was a very basic tractor, designed and made by RA Lister at Dursley, Gloucs, for the overseas market. Introduced in 1958 it had a twin-cylinder air-cooled Lister diesel engine, a six forward and one reverse gearbox, conventional final drive and a drawbar. There was no electrical system and the tractor was started by hand with a foldaway handle. A self-bleed system was provided to get the engine started if it ran out of fuel. The front wheel track was adjustable in 4 in steps from 44 to 64 in and the rear wheels from 40 to 68 in. The absence of a water radiator facilitated the Goldstar's modern-looking low-profile bonnet which provided good visibility for the driver.

Power output was similar to that of a Fordson Dexta or Massey Ferguson 35 but the Goldstar's specification was inferior to these tractors, and to improve sales an optional electric starting and lighting kit was introduced. Even then the planned volume production of the tractor failed to materialise.

MARTIN-MARKHAM COLT

About 100 Colt garden tractors were made between 1961 and 1970 by Martin-Markham at the Lincolnshire Ironworks in Stamford and marketed by Colt Tractors of the same address. The Martin Cultivator Co, established at Stamford in the early 1900s, merged with the Stamford trailer manufacturer Markham Traction Ltd in 1952. Although better known for its farm trailers Martin-Markham also made rotary cultivators, forage harvesters and barn machinery.

The Stamford company was taken over in 1970 by Spiroflite which made grain augers but the new owners went into receivership in 1972. As no buyer was forthcoming the business closed down and the premises were eventually demolished to make way for a retail park.

The metallic blue Colt tractor with red wheels was usually built in small batches of six, when space permitted at the Lincolnshire Ironworks. With the exception of the engine and hydraulic pump most of the components were also made on the premises. The standard Colt tractor had an air-cooled Kohler 7 hp four-stroke engine with a recoil starter. A 12 volt starter motor, dynamo, battery and lights were optional extras.

Engine power was transmitted by flat belt to a single-plate clutch and then by a shaft to a transfer box. From here the drive was by roller chain to a three forward speed and one reverse gearbox on the rear axle. The specification included drop arm worm-

2.43. The Martin-Markham Colt de luxe was advertised as the most practical and versatile tractor of its class. (Roger Smith)

and-ball steering, a fixed drawbar and a rather ineffective internal expanding shoe brake on one rear wheel.

The rear wheel track setting was adjustable in three 4 in steps from 32 to 40 in and the front track from 29 to 37 in. Front, centre and rear power take-off

2.44. Martin-Markham made a front-end loader for the Colt tractor.

shafts were optional. With the engine running at 3,000 rpm rated speed the rear power take-off ran at 280 rpm while the centre and front shafts turned at 1,500 rpm.

A 6 in diameter belt pulley could be used on the centre power shaft and the platform at the front provided a convenient space to mount and drive an air compressor or an irrigation pump. A gear pump, with an output of four pints per minute, supplied oil to the ram cylinder used for the live four-point hydraulic linkage and also to an external ram when the Colt was used with a small tipping trailer or front-end loader.

A 10 hp air-cooled Kohler four-stroke engine with electric starting was the most significant improvement on the Martin-Markham Colt de luxe which cost £350 ex works when it was introduced at the 1965 Royal Smithfield Show. The de luxe tractor had a new constant mesh gearbox, six forward gears with top speeds between 1½ and 6 mph and two reverse gears. Braking efficiency was improved with independent internal expanding shoe brakes on both of the rear wheels and there was also a parking brake. It had a conventional rear power take-off with a British Standard six-spline shaft tuning at 540 rpm.

An optional front power take-off shaft running at 1,000 rpm with the engine at its 2,400 rated speed was used to drive the live hydraulic pump.

The adjustable wheel track and fixed drawbar were identical to those on the standard tractor but instead of a steel pan seat the Colt de luxe driver enjoyed the luxury of a padded seat with a curved back rest. Sales literature described the Colt de luxe as the only small tractor with built-in live four-point hydraulics and two power take-off shafts, making it the most versatile tractor in its class. Implements for both Colt tractors included the usual selection of ploughs and cultivation equipment along with a front-end loader, a tipping trailer, a rotary grass cutter, a forklift attachment, a dozer blade and even a concrete mixer.

MASKELL

In 1961 H Maskell & Son of Wilstead near Bedford introduced the three-wheel Maskell rowcrop tractor at the National Spring Sugar Beet Demonstration. It had a 14 hp Enfield 100 twin-cylinder air-cooled diesel engine, a heavy-duty single-plate clutch and a three forward and one reverse gearbox. A reduction

2.45. The Maskell rowcrop tractor was made in Bedfordshire in the early 1960s.

differential eliminated the need for a crown wheel and pinion and a roller chain transmitted drive to the rear wheels.

The Maskell had independent rear brakes and the single or optional twin front wheel was steered with a tiller handle. The optional hydraulic system had a 15 gallon rear-mounted oil reservoir and the engine-driven pump with an output of 8½ gal/min supplied oil to the hydraulic linkage ram and to a hydraulic motor which could be attached at various points on the hollow rectangular steel chassis. The Maskell had a minimum 17 in ground clearance, a fuel consumption of less than a gallon an hour and a top speed of 8 mph.

Barfords of Belton, part of the Aveling Barford Group, made the modified Barford Maskell rowcrop tractor in the mid-1960s. When compared with the low-level tool frame of the earlier model the Barford-Maskell had a high-arched, hollow rectangular section steel open chassis. An engine-driven pump supplied oil to a double-acting ram used to raise and lower the toolbar pivoted on the front end of the chassis. Toolbar equipment included various attachments for inter-row cultivations, a set of seeder units driven by a hydraulic motor and a sprayer.

2.46. The Barford-Maskell rowcrop tractor.

MASSEY FERGUSON

Massey Ferguson was not involved with garden tractors until 1965 when the 10 hp MF10 Suburban tractor was introduced in North America. Designed for market gardens and large domestic gardens, the MF10 had a single-cylinder four-stroke air-cooled

2.47. The Massey Ferguson 1010 compact tractor.

petrol engine. Gear changes were made on the move through a five-speed vee-belt drive combined with a four forward and one reverse gearbox which gave a total of twenty forward gears and five in reverse. The 7 hp MF7 lawn tractor and the 12 hp MF12 with hydrostatic transmission were added a year or so later.

A full range of cultivators, mowers and other implements including a trailer and a snow blower were made for these tractors. They were sold in continental Europe and were particularly popular in France. They were made until 1977 when an agreement with Toyota resulted in the launch of a new range of Japanese-built Massey Ferguson 1000 series compact tractors.

Although it was sold in some parts of Europe the 1000 series tractors was not available in the UK until 1983 when the MF1010 was exhibited at the Royal

Agricultural Show. The MF1010, along with the MF1020 and MF1030 compact tractors, were introduced to the British market in May 1984. All three were available with two- or four-wheel drive and had three-cylinder water-cooled diesel engines. The smallest 16 hp MF1010 with six forward and two reverse gears, power take-off and category I hydraulic linkage was suitable for the market gardener.

The 21 hp MF1020 and 27 hp MF1030, both with a twelve forward and three reverse gearbox, including a range of creep speeds, were intended for specialist horticultural planting and harvesting operations. An optional hydrostatic transmission with a high/low ratio lever and a rocker pedal to control forward and reverse travel was introduced in 1987.

The MF 30 series lawn tractors, launched by Massey Ferguson in 1990, had a single lever clutchless speed control for forward and reverse and four-wheel

2.48. The Massey Ferguson 1020 Hydro compact tractor.

steering to give a tight turning circle. The three models, designated the MF 30-13, MF 30-15 and MF 30-17, had hydrostatic transmission and mid-mounted side-discharge rotary mower decks. The second pair of figures in the model number denoted the horsepower developed by their Briggs & Stratton Vanguard V-twin petrol engines.

The MF 20-12 with a single-cylinder Briggs & Stratton power unit and a six forward and one reverse manual gearbox completed the 1990 range of MF lawn tractors. All four had a drawbar for a small trailer and trailed turf care equipment.

The Massey Ferguson 12 Series, introduced in 1993, was made by Iseki in Japan. The MF 1210, MF 1220 and MF 1230 were compact models with 17, 20 and 25 hp diesel engines and the 30 hp MF 1250 and

35 hp MF 1260 tractors were for golf courses, parks departments, etc. Options for the 12 series included mechanical and hydrostatic transmissions, manually engaged front-wheel drive and power steering.

MOUNTFIELD

In the late 1970s G M Mountfield of Maidenhead made its own garden and lawn tractors and also imported Wheel Horse D series tractors from America. The 11 hp Mountfield garden tractor and 10 hp Mountfield lawn tractor had Briggs & Stratton engines, a single-plate clutch and a three forward and one reverse gearbox.

In the mid-1980s Mountfield lawn tractors included the Briggs & Stratton-engined 10 hp 10/30 with a five forward and one reverse gearbox and the

11 hp 11/40 lawn tractor with four forward gears and one reverse gear. Also made at Maidenhead at that time were the 8 hp Mountfield 8 and 10 hp Mountfield 10 lawn tractors with a three-speed transaxle. At the top of the range was the 16 hp Mountfield 16 garden tractor with a twin-cylinder Briggs & Stratton engine and four-speed gearbox.

G D Mountfield was selling the American-built Simplicity lawn and garden tractors in 1988 along with the renamed Mountfield Murray 10/30 and 12/40 lawn tractors. The 10/40 was unchanged but the 11/40 with an extra horsepower under its bonnet became the 12/40. Like other Mountfield tractors it had a drawbar for towing a trailer, a lawn sweeper, an aerator and other trailed equipment.

2.49. The Mountfield 8 had a Briggs & Stratton four-stroke engine, three-speed transmission and a tow bar for a sweeper, roll and dump truck.

MSW

MSW Machinery of Wood Vale in London bought the manufacturing rights and remaining stocks of Winget tractor parts from Slater & England in 1968 and made MSW tractors until the late 1970s. When the stock of Winget parts was used up MSW Machinery bought in alternative components including a Newage gearbox. The troublesome Winget transaxle was replaced with a conventional rear axle unit.

The basic specification of the early 1970s MSW 550 four-wheel drive tractor included a 9 bhp single-cylinder Lister SR1 four-stroke diesel engine, a single-plate clutch and a combined gearbox and rear axle with three forward speeds and one reverse speed. The tractor also had a diff-lock, independent rear wheel brakes and adjustable wheel track settings. An independent power take-off and a live hydraulic system with an engine-driven pump were standard but a hydraulic trailer pipe and electric starting were optional extras.

2.50. The MSW 550 compact tractor.

Four MSW models were listed in 1973. The MSW 550 had become the MSW 550M to denote its manual-change gearbox and the MSW 550HC with a Petter BA 1 diesel engine had the Marshallmatic fully variable forward and reverse hydrostatic transmission. The MSW 633M also came with a manual-change gearbox and a 9 hp Lister SR1 or an 11 hp Lister ST1. The MSW 633HC with the Lister ST1 engine and Sandstrand 18 fully variably hydrostatic transmission completed the range.

Variants of the MSW 633 had a heavy-duty rear axle and alternative Kohler petrol or Hatz diesel engines were available on special order. Prices ranged from £891 for the MSW 550M to £1,084 for the MSW 633 HC with an 11 hp Lister diesel engine. There were six models in 1975 when the 19 hp Lister SR2 diesel-engined MSW 1250M and MSW 1250HC with a Sandstrand hydrostatic transmission had been added to the price list. Matched equipment made for MSW tractors included a front-end loader, ploughs and cultivators, trailer, mower, sprayer and an irrigation pump.

NASH ROLLER TRACTOR

Sales literature described the pillar-box red Nash Roller Tractor as a multi-purpose power unit and a remarkable maid of all work designed for mechanical handling, towing, rolling and mowing. The three-wheel Nash Roller tractor made by H R Nash at Dorking was launched at the 1950 Smithfield Show. It was advertised as a garden tractor, tipper dumper and mini-roller designed for use on market gardens, poultry farms, sports grounds and, to a lesser extent, on building sites.

The Nash had a full-width driven roller between the rear wheels and it took less than a minute to remove the wheels and convert the machine to a power-driven roller for lawns, cricket pitches and other sports ground facilities. Consolidation could be increased with ballast weights in the hopper, which was also used to transport liquid and solid loads. The roller was also said to serve as a flywheel for the Coventry Victor engine and its weight improved traction when travelling across rough ground.

The Nash Model A had a hand-started 3½ hp Coventry Victor air-cooled flat-twin petrol engine and the later Nash Model B with a 5 hp BSA G series air-cooled single-cylinder engine had a power take-off shaft for driving a saw bench and other equipment. Engine power on both models was transmitted by roller chain to a three forward and one reverse Albion gearbox. A second roller chain linked the gearbox output shaft to the rear wheels giving forward speeds of up to 18 mph.

The Nash was steered with a tiller handle while other driving controls included a lever-operated clutch, rear wheel brakes and a pedal lock for parking. The Roller Tractor had a 4 cu ft capacity tipping hopper, and towing an optional two-wheel trailer increased the load capacity. Rear-mounted or trailed implements included a toolbar with cultivating tines, ridging body, potato lifter, disc harrow and a Shanks 30 in mini-gang mower. A front-mounted toolbar was made for rowcrop work.

The 600 Group acquired H R Nash in 1953 when George Cohen & Sons was appointed sole distributor for Nash Roller Tractors in the UK. The four-wheel Nash 10 Roller Tractor dumper introduced in the same year cost £280 and apart from the extra front wheel, car-type steering wheel, 10 cwt tipping hopper and the ability to run on petrol or tvo, was similar to earlier machines.

Landmaster at Hucknall was selling Landmaster-

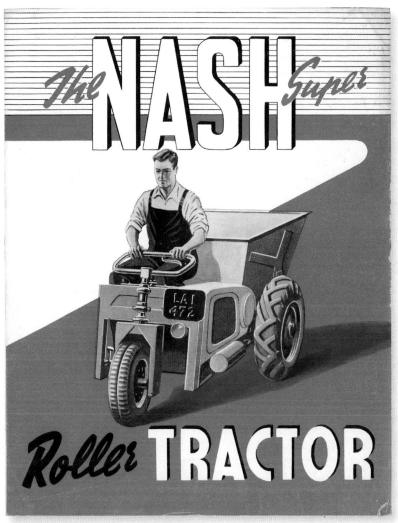

2.51. The 1950s Nash Roller tractor was a dump truck with a difference in that with the rear wheels removed it became a road or land roller.

Nash Roller tractor-dumpers in 1955. The Nash 12 four-wheel dumper without rollers had a 12 cu ft tipper hopper and the choice of a 3½ hp Lister diesel or 5 hp Petter diesel engine. The Nash 10 four-wheel roller tractor-dumper with a 10 cu ft capacity tipper hopper and the three-wheel Nash 8 tractor-dumper 10 cu ft capacity tipper hopper both had a 5 hp petrol or tvo engine. Car-type steering with a wheel, mudguards and a towbar were standard and twin rear wheels were an optional extra for the Nash 8 and 10 roller tractors. Attachments included a manual lift rear toolbar, hedge cutter, sprayer, light triple gang mowers, a trailer and a weather cab.

NEWMAN

The Grantham area of Lincolnshire had more than its fair share of small tractor manufacturers in the late 1940s. Grantham Productions, which manufactured the three-wheel Kendall tractor from 1945 to 1947, was bought by Newman Industries of Yate near Bristol and the new owners made the first three-wheel Newman light tractors at Grantham in 1948.

Air-cooled horizontally opposed Coventry Victor petrol engines were used for the 10 hp Newman AN3 and the 12 hp Newman AN4 which was recommended for districts where more power was needed to deal with difficult soil conditions. Otherwise both Newman tractors were identical with a single-plate dry clutch, a three forward and one reverse gearbox with a top speed of 10 mph, independent rear wheel brakes and swinging drawbar. Optional equipment included an underslung toolbar with hydraulic or manual lift, rear toolbar, power take-off and a belt pulley.

2.52. Following the acquisition of the Kendall tractor business, Newman Industries introduced the three-wheeled Newman tractor in 1948.

The Newman WD2 introduced in 1949 had a water-cooled single-cylinder Coventry Victor diesel engine. Two pipes connected the mid-mounted engine to the radiator and cooling fan at the front of the tractor. The tractor had a single-plate dry clutch and a conventional transmission system. With the engine at full throttle the gearbox provided top speeds of 1½, 3 and 9 mph and 1 mph in reverse.

There were independent rear wheel brakes and by sliding the wheels along the axles it was possible to set the

2.53. A lever-operated manual implement lift was included in the price of the Newman WD2.

rear track width anywhere between 42 and 72 in. A lever-operated chain lift was used to raise and lower the mid- and front-mounted toolbars and rear-mounted implements. The WD2 cost £330 in 1949 but with a belt pulley, power take-off and hydraulic linkage the total cost was £410 12s 0d, making the Newman more expensive than a Ferguson TEF 20.

The four-wheel Newman Model E2 tractor with four forward gears, one reverse and a top speed of 8¼ mph was introduced in 1951. The front wheel track was adjustable from 42 to 54 in and the rear wheels could be set to a maximum of 52 in. The tractor had a 12 hp Petter AV2 twin-cylinder water-cooled diesel engine started by hand with the aid of a decompressor and a fuel priming lever.

The basic tractor cost £430. Extras included a 9 in diameter rear belt pulley for £10, a 1⅛ in power take-off shaft for £18 and a 3½ ton capacity hydraulic linkage with a gear pump in the transmission housing which added £46 10s 0d to the price. The E2 had a swinging drawbar. The Newman's high ground

clearance and mid-mounted toolbar made it an ideal tractor for rowcrop work.

OPPERMAN

The three-wheel farmer-designed Motocart and four-wheel Tractivator were made by S E Opperman of Stirling Corner at Boreham Wood in Hertfordshire in the 1940s and 1950s.

Jack Olding & Co at Caterpillar Island in Hatfield was the sole concessionaire for the three-wheeled Opperman Motocart. An advertisement in 1946 suggested that with a top speed of 11.2 mph and fuel costs of only 2s 6d a day the Motocart was the most economical 30 cwt transport vehicle available.

The four-stroke air-cooled 8 hp engine with a Zenith carburettor and a Wico magneto was attached to the offside of its large-diameter single front wheel and connected by roller chain to a single-plate clutch, a four forward and one reverse gearbox and spur final reduction gears. The Motocart had

2.54. The Opperman Motocart.

2.55. The Opperman Tractivator.

ensured that the tines and hoe blades followed the ground contours without the need for depth wheels. Left- and right-handed plough bodies, raised and lowered hydraulically, were made for the Tractivator which could also be used for ridging and other rowcrop work. The wheel track was adjustable in 4 in steps from 38 to 72 in. According to the manufacturers, one gallon of petrol was sufficient to keep the tractor working for three hours.

OTA

Oak Tree Appliances of Coventry used the company initials for the three-wheel OTA light tractor which, after an appearance at a local agricultural event in 1948, was launched nationally at the 1949 Royal Smithfield Show. Oak Tree Appliances and Slough Estates of Berkeley Street in London, the sole concessionaire for the OTA, introduced the four-wheel OTA Monarch in 1951 when it cost £297 10s 0d compared with £269 for the three-wheel tractor.

expanding shoe brakes on the rear wheels and the option of a rigid or screw-tipper wooden body. The patent steering wheel could be set in different positions, allowing the Motocart to be steered from the driving platform or while walking alongside the machine.

In 1948 a prototype Tractivator was exhibited at the Royal Agricultural Show. Within a couple of years the Opperman Tractivator was in production and cost £249 complete with pneumatic tyres and hydraulic linkage. A mid-mounted toolbar with hoe blades and cultivator tines added £28 10s 0d to the price.

The Tractivator had a channel section steel chassis and a 5 ft 10 in wheelbase. Power from the rear-mounted Douglas 8 hp four-stroke engine was transmitted by roller chain through a single-plate clutch to a four forward and one reverse gearbox. The clutch pedal and independent brake pedals were attached to the chassis side members, while the steering wheel and gear lever were in a central position in front of the driver.

An engine-mounted pump supplied oil to a hydraulic lift ram for the mid-mounted toolbar attached by a parallel linkage arrangement to the front axle. This

The first OTA tractors were red and yellow but most were painted blue and this colour was still used

2.56. The OTA tractor was made at Coventry.

in 1953 when the manufacturing and marketing rights for the OTA tractor were sold to the Singer Motor Co. The new owners discontinued the three-wheel model and transferred production of the Monarch to the Singer car factory at Birmingham. A total of about 550 OTA tractors, including 293 Singer Monarch tractors, had been made when the Rootes Group acquired Singer Motors in 1956.

The three-wheel OTA had a water-cooled side-valve 10 hp Ford industrial petrol engine with an optional Beccles Engineering Co vaporising oil conversion kit. The electrical system consisted of a 6 volt battery, generator, starter motor and coil ignition, but lights were an optional extra.

In those days it was usual to provide a starting handle. An unusual feature of the standard Mk I tractor was the provision of a starting handle dog on the crankshaft pulley and a second dog at the end of the shaft carrying the high/low ratio gears at the back of the gearbox. As the bottom of the engine radiator obstructed the starter dog on the crankshaft pulley the OTA could only be started from the rear with the main lever set in gear and the high/low ratio lever in neutral.

The OTA had a Ford gearbox which, combined with a high/low ratio box, provided six forward and two reverse gears with a top speed of just under 4 mph in low agricultural range and 13½ mph in the high industrial range. The transmission system comprised a worm-and-wheel final drive, differential and internal expanding shoe brakes. The chassis consisted of two steel channels set at an angle to raise the front of the tractor where a fork assembly carried the cable-steered front wheel which gave the OTA a tight turning circle.

A small range of implements for the Mk I OTA included a single-furrow plough with a Fisher Humphries body, a mid-mounted toolbar, a combined four-speed power take-off shaft and a belt pulley. An optional live hydraulic system which had a twin-cylinder piston pump belt-driven from the engine crankshaft pulley was added at a later stage. The tractor's high ground clearance gave the driver an excellent view of the mid-mounted toolbar when working in rowcrops.

The most obvious change to the Mk II OTA introduced a year or two later was the redesigned bonnet and sheet metal radiator grille. Some Mk II tractors, also known as the 5000 series, had an improved single unit transmission housing instead of separate castings.

The four-wheel Monarch launched at the 1951 Smithfield Show was, apart from the obvious change to the front axle with orthodox steering linkage and a front-hinged bonnet, almost identical to the tricycle tractor. Wheel track settings were adjustable from 42 to 60 in and various hydraulic-operated implements included mid- and rear-mounted toolbars, plough, spike tooth harrow, disc harrow, tined cultivator, potato lifter and a mid-mounted mower.

The OTA Monarch became the Mk III Singer Monarch in 1953. An improved Mk IV Monarch with a new orange colour scheme and standard category I three-point linkage appeared in 1955. The Monarch's five-year production run came to an end in 1956 when Singer Motors became part of the Rootes Group.

RANSOMES

In 1932 Ransomes built and tested a prototype pedestrian-controlled petrol-engined garden tractor with rubber-jointed tracks made by Roadless Traction, but the project was soon dropped. Ransomes then turned its attention to building an experimental ride-on garden tractor with Roadless rubber jointed tracks.

The Ransomes Motor Garden Cultivator was demonstrated for the first time in Worcestershire 1936 when the Roadless Traction News explained that it was the first ride-on tractor to be designed for the smallholder, market gardener and fruit farmer and could do the work of two horses.

The Ransomes MG (Motor Garden) Cultivator, later known as the MG2, had a 600 cc air-cooled, single-cylinder Sturmey Archer 'T' petrol engine, one forward and one reverse gear and 6 in wide Roadless rubber-jointed tracks. The 6 hp side-valve engine had a dry sump with a separate tank for the lubricating oil, which was pumped to the bearings and then returned by a second pump through a filter to the tank. It had a Lucas magneto without an

2.57. The Ransomes MG 2 was built at Ipswich from 1936 to1948. (Roger Smith)

under some degree of power to eliminate slewing or sliding when changing direction.

The centrifugal clutch acted as a safety overload mechanism, so if the tractor was overloaded the engine speed fell away and the clutch disengaged the drive. The tracks had a ground pressure of about 4 psi and the track width was adjusted by fitting different-sized spacing blocks between the track mounting points and the chassis to give three settings between 2 ft 4 in and 2 ft 10 in. Gears replaced the chain drive from the starter dog to the crankshaft on the improved Sturmey Archer 'TB' engine which replaced the 'T' power unit in 1938. It had a belt-driven cooling fan and with an impulse coupling in the Wico magneto it was easier to start the engine.

Ransomes' sales literature explained that the MG2 could plough an acre with a single-furrow plough or cultivate five acres with a seven-tine mounted cultivator in an eight-hour day. The MG2 with a hand-lift tool frame and swinging drawbar cost £135 in 1936, the optional 400 rpm power take-off shaft adding £1 10s 0d to the price. A considerable number of MG2s were used in French vineyards and Ransomes was astute enough to provide both imperial and metric dimensions and capacities in its sales literature.

impulse coupling and the starting handle dog was attached to a countershaft connected by a roller chain to the crankshaft. Power was routed through a 4:1 reduction gearbox on the engine output shaft to a centrifugal clutch that automatically engaged the drive to the tracks when engine speed reached approximately 500 rpm.

The MG2 had a top speed of about 2 mph in both directions and a single lever was used to select forward, neutral or reverse. The transmission consisted of an inward-facing pair of crown wheels and a driving pinion on the clutch output shaft. The lever was used to engage the pinion with one of the crown wheels to select forward or reverse. It was steered with two lever-operated band brakes on the drive shafts to the tracks and both shafts were always

About 3,000 MG2 tractors were produced between 1936 and 1948 at the Long Street Works in Ipswich and, despite material and labour shortages, about 1,200 were made during World War II. Implements for the MG2 included the TS25 one- and two-furrow hand-lift ploughs, disc harrows and a mounted tool frame with cultivator tines, hoe blades and ridging bodies. The TS42 self-lift single furrow plough with optional subsoiling tine was added a year or so later.

In 1949, in anticipation of the possible exhaustion of fossil fuels, the Electrical Research Association decided to build an experimental electric tractor. An MG crawler with a 9 hp electric motor was used for the purpose with the power supplied by cable from a 35 ft high pylon. The cable was attached to a rotary

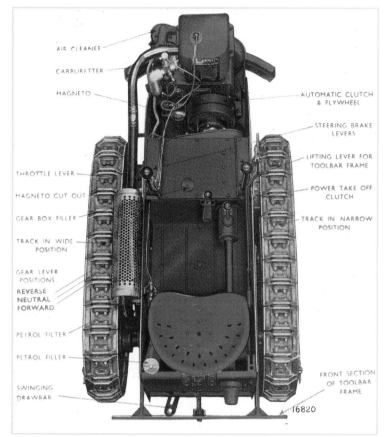

2.58. Sales literature claimed that the MG5 was so simple to operate that a boy could use it with ease.

output shaft, forward and reverse gearbox were retained and the hand lift toolbar and drawbar were included in the basic price of £250.

Optional extras included a vaporising oil conversion kit and a hydraulic lift unit made by R J Neville in Australia. An increased range of implements for the MG5 included an improved TS42A single-furrow plough, HR4 disc harrows, a C29 toolbar with cultivator tines, hoe blades, ridging bodies and a potato-raising plough. Orchard spraying equipment made by Coopers of Wisbech and a small trailer were among a list of approved implements made by other companies.

The 600 cc Ransomes side-valve petrol or two engine and centrifugal clutch were retained on the MG6, launched at the 1953 Smithfield Show. However, the new model had a three forward and three reverse gearbox with top speeds of 1, 2¼ and 4 mph. An alternative 8½ hp Drayton two-stroke, overhead-valve air-cooled engine diesel engine became available in 1956. The hydraulic linkage and 700 rpm power take-off were still optional.

connector which allowed the tractor to turn through 360 degrees and work within a 110 ft radius round the pylon. Counter balance weights on the pylon kept the cable under tension while the tractor was at work.

The MG5, which superseded the MG2 in 1948, was designed to do two-horse work at two-horse speed and was advertised as the complete answer to the mechanisation of holdings of up to 25 acres. The power take-off speed was increased to 700 rpm but the more obvious differences were the fuel tank located under the seat and a cowling over the 600 cc Ransomes air-cooled petrol engine. The dry sump lubrication system, 4:1 reduction gears on the engine

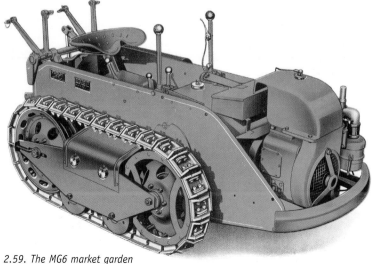

2.59. The MG6 market garden cultivator was launched in 1953.

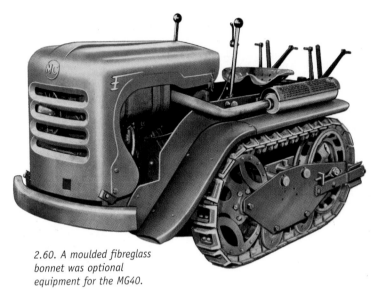

2.60. A moulded fibreglass bonnet was optional equipment for the MG40.

mounted dozer blade. An alternative version had modified steering levers and, with the driver sitting with his back to the engine, the ITC could be driven in reverse with a digger-loader or a dump hopper mounted on what would normally be the back end of the tractor.

When Ransomes discontinued the ITC and ITW, a limited number of tracked WR4 dumpers and WR8 loaders based on the MG6 and later on the MG40 were made by Whitlocks of Great Yeldham in Essex.

The MG40, which superseded the MG6 in 1960, was painted Ransomes lawn mower green. There was a choice of a petrol, tvo or diesel engine with wet sump lubrication. The 8 hp four-stroke side-valve petrol and petrol and tvo engines had an Amal carburettor and Wico magneto. The 10 hp two-stroke overhead-valve diesel engine, started from cold with an ignition wick, had a fuel consumption of about three pints an hour. The MG40 had the same centrifugal clutch, three forward and three reverse gearbox and differential spur gear reduction units as the MG6.

Other extras for the MG6 included a parking brake, a belt pulley, a pair of front-steadying wheels, detachable rubber track pads and 8, 10 or 12 in wide hardwood track blocks for boggy or swampy land. The front-steadying wheels were used to stop the front of the tractor 'digging in' on steep downhill slopes or undulating land. Detachable rubber pads were recommended when using the MG6 on hard roads or inside buildings where the tracks might damage the surface.

Industrial Tractor Wheeled (ITW) and Industrial Tractor Crawler (ITC) versions of the MG6 appeared in 1956. Heavy-duty roller chains were used to transmit drive from the standard crawler transmission to the front wheels on the skid-steered ITW wheeled model which was mainly used for haulage work and shunting railway wagons.

The ITC tracked model had the same track layout as the standard MG 6. Optional rubber blocks could be bolted to the track plates when working on hard surfaces and, when equipped with hydraulic linkage, it could be used with a front-

2.61. Ransomes compact tractors were made in Japan.

Improvements in 1962 included new steel track guards, needle roller bearings in the track rollers and idler wheel hubs and an optional moulded fibreglass bonnet. Later MG40 crawlers had a Sachs diesel engine. About 15,000 MG crawlers, including 3,000 MG2s, 5,000 MG5s and MG6s and about 2,000 MG 40s had been made when production ceased in 1966.

Ransomes returned to the tractor market in 1997 when it introduced a range of Japanese-built four-wheel drive compact tractors. The CT318, CT320 and CT325 had 18, 20 and 25 hp three-cylinder diesel engines with a manual or hydrostatic transmission. The 33 hp three-cylinder CT333 HST had a three-speed hydrostatic transmission with cruise control and an independent hydraulic system to raise and lower its mid-mounted mower deck. The 38 and 45 hp four-cylinder diesel-engined CT435 and CT445 had a shuttle transmission, mid and rear power take-off shafts, power steering and a quiet cab.

ROLLO CROFTMASTER

Rollo Industries, based at the Barrmor Tool Works in Bonnybridge, Scotland first made the four-wheel Rollo Croftmaster in 1953. It was designed by Scottish crofter John Rollo for work on 5 to 10 acre holdings in the highlands. Known as the Mk 6 it had a box section steel chassis with a chain drive from the Mk 20 Villiers engine to a hand lever operated clutch.

Power was transmitted by a three forward and one reverse Albion gearbox with a top speed of 6 mph and a countershaft to the differential and rear axle. A pedal-operated Dunlop expanding shoe brake on the differential shaft stopped both wheels, while the foot pedal was backed up by a hand lever which could be used when it was desirable to stop in a hurry.

A parallel linkage toolbar, designed to keep the implements level with the ground at all times, was raised with a hand-operated hydraulic pump on the right-hand side of the tractor. Oil for the toolbar lift ram was drawn from a separate tank and six to eight strokes of the pump lever were sufficient to lift a single-furrow plough well clear of the ground. A tap was used to release the oil from the ram cylinder and lower the implement.

Later Mk 6 Rollo Croftmaster tractors were sold with a 3 hp BSA engine but lacked power. It was replaced by a 5 hp BSA engine, some of which were supplied with a two conversion kit.

A 5 bhp JAP single-cylinder engine was used for most Mk 7 Croftmaster tractors when they were introduced in 1955, although some had a 3 bhp Villiers Mk 25 and a few, mainly for export, had the more powerful Villiers Mk 40 engine. The Mk 40 was considered more suitable for heavy work or for use at high altitudes where there might be a drop in power output.

The Mk 7 Rollo was similar in appearance to the earlier model but a new Albion gearbox with a five-plate clutch and a chain-driven reduction box with a built-in differential facilitated the use of separate drive shafts to the rear wheels. A more efficient transmission brake was linked to the reduction gearbox and oil from the gearbox was used for the hand-operated hydraulic lift unit.

2.62. The Mk 6 Rollo Croftmaster.

The 3 hp Mk 7 Croftmaster with a toolbar and hand-operated hydraulic unit cost about £315, the 5 hp engine adding another £13 to the price. The single-furrow plough was £18 10s 0d, a 3 ft motorised cutter bar mower cost £66 and a 6 cwt two-wheel trailer was £34 10s 0d.

Introduced in 1957, the Rollo Croftmaster 7C was mainly used by local authorities and on larger holdings. It had a much heavier chassis and was known in some circles as the Big Yin, or Big One. A 9 bhp Briggs & Stratton engine provided the power, the standard Mk 7 transmission was retained and it had parking brakes on both rear wheels. A hand-operated twin-cylinder hydraulic pump took its oil from the gearbox and the tractor was styled with sheet metal covers over the gearbox and back end. A cushioned seat and backrest made life a little easier for the driver.

RUSSELL

There was still sufficient demand in the mid-1970s for Russells of Kirbymoorside, Yorks, to manufacture the Russell 3-D powered tool carrier. It had a rear-mounted twin-cylinder air-cooled 20 hp diesel engine with electric starting. A single lever controlled the infinitely variable hydrostatic transmission with a hydraulic motor in each rear wheel with a maximum forward and reverse speed of 9 mph. Separate levers were used to raise and lower the mid-mounted toolbar and rear-mounted wheel track eradicators.

The Russell 3-D had independent disc brakes and the wheel track was adjustable from 48 to 76 in. Cultivator tines, hoe blades and seeder units were among the range of attachments for the 3-D which cost about £4,600 in 1978, rising to £4,950 by 1985.

2.63. The Russell 3-D self-propelled tool carrier was made in the mid-1950s.

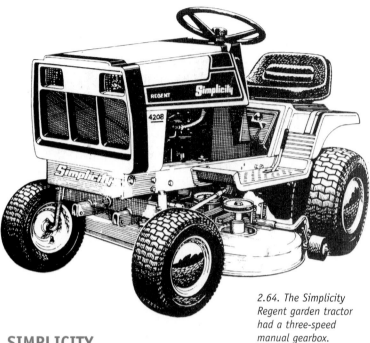

2.64. The Simplicity Regent garden tractor had a three-speed manual gearbox.

SIMPLICITY

Stemport Marketing Co, based at Aylesbury, imported Simplicity garden tractors from America in the late 1970s and early 1980s. Most Simplicity tractors could be used with a mid-mounted rotary lawn mower and other attachments. Simplicity models in the late 1970s included the 8 hp and 10 hp Broadmoor, the 8 hp Regent, the 10 hp Landlord, the 13 hp Baron and the 16 hp Sovereign. Some had hydrostatic transmission, others were equipped with a manual gearbox and most had an electric starter.

Al-Ko Britain Ltd, based at Consett in County Durham were marketing Simplicity garden tractors in the mid-1980s when the much reduced range consisted of the 8 hp Cavalier and the 12 hp Regent with the choice of a manual or hydrostatic transmission. G D Mountfield sold 12½ and 16 hp Simplicity garden tractors with

Briggs & Stratton engines in the late 1980s until Simplicity, an Allis Chalmers company, established its own Simplicity Manufacturing (UK) outlet at High Wycombe in the early 1990s.

The 1993 Simplicity tractor range included the 12½ hp Regent and Broadmoor lawn tractors as well as the 14 or 18 hp Landlord and 20 hp Sunstar with Kohler engines, hydrostatic transmission and hydraulic linkage.

SINGER MONARCH

Singer Motors, which acquired the manufacturing and marketing rights of the OTA Monarch, started production of the four-wheel tractor at its Birmingham car factory in 1953. Renamed the Singer Monarch Mk III, it had the same four-cylinder Ford 10 industrial engine with a Beccles Engineering Co tvo conversion kit and a 6 volt electrical system.

2.65. The Singer Monarch had a fuel consumption of about half a gallon an hour.

2.66. The Stockhold President with a Hayter mower.

The specification included a single-plate dry clutch, a six forward and two reverse gearbox, hydraulic linkage, four-speed power take-off with a belt pulley and a 42 to 60 in adjustable wheel track. Singer Motors exhibited the Monarch at the 1953 Royal Smithfield Show when it was advertised as the tractor which small and medium farmers everywhere should consider first for all types of arable work.

The Singer Mk IV appeared in 1955. Apart from a colour change from blue to orange paint work the most significant improvement was the inclusion of the category I hydraulic linkage within the standard price of £279 ex works. Mounted implements made for the Monarch included a plough, tined cultivator, disc harrow, steerage hoe, mower and a potato lifter.

A total of about 550 OTA and Singer Monarch tractors including a few Mk IV Monarchs had been sold when production came to an end following the acquisition of Singer Motors by the Rootes Group in 1956. A firm of agricultural engineers in

Birmingham bought the remaining stock and spare parts and then sold them as replacement parts to existing owners of Monarch tractors.

STOCKHOLD PRESIDENT

Following the demise of the BMB President in 1956 H J Stockton of London decided there was still a viable market for a small light tractor which was little more than an engine, transmission, drawbar and seat on four pneumatic-tyred wheels.

The Stockhold President was exhibited at the 1957 Royal Smithfield Show. It was based on its BMB namesake and similar in appearance with the main change being the twin-cylinder air-cooled 14 hp Petter diesel engine. The price for the basic tractor including a starting handle was expected to be about £360, with power take-off, hydraulic linkage and lights available at extra cost. However, sales of the Stockhold President did not meet expectations and very few were made.

2.67. The Trojan Monotractor was introduced in 1963.

TROJAN

The three-wheel Monotractor, manufactured by Trojan Ltd at Croydon in Surrey, was designed by the National Institute of Agricultural Engineering mainly for the overseas market but it was also sold in the UK as a rowcrop and runabout tractor. In 1963 the export model cost £185 but for the home market the price was £270. A mid-mounted toolbar was an extra £12 10s 0d and the optional hydraulic unit for the front toolbar and other heavy front-mounted attachments added £120 to the price.

The Monotractor had a single driving wheel at the rear. The two front wheels were tiller-steered and a fourth depth wheel was used when doing rowcrop work with the front toolbar. The specification included a 7 hp Clinton or Villiers four-stroke engine, an over-centre flat-belt clutch and a two forward and one reverse gearbox.

Equipment for the Monotractor included a plough, ridger and Webb seeder units for the front toolbar. There were also a mid-mounted plough as well as a Dorman low-volume sprayer with a 30 gallon front tank and an 18 ft spray bar at the rear. A 3 ft 6

in cutter bar mower with its own engine and a front load platform were also made for the Monotractor.

The four-wheel Trojan Toractor designed to provide owners of large gardens with a mini-tractor for mowing, transport and cultivating was made in the mid- to late 1960s. The blue Mk I had pneumatic tyres and an air-cooled Clinton engine mounted crosswise on the chassis.

It was followed by the red and black Mk III marketed by the Toractor Co in Godalming, Surrey, and later by Trojan Agricultural Sales at Croydon. It had a conventionally mounted air-cooled single-cylinder 3½ hp Clinton engine with a top speed of 4 mph and 1½ mph in reverse. It cost £145 ex works in 1967. A rotary cultivator with a separate 4 hp Clinton engine to drive the cultivating rotor was one of several attachments made for the Toractor.

TRUSTY STEED

Tractors (London) Ltd, based at Bentley Heath, already famous for the two-wheel Trusty garden tractor, exhibited an experimental ride-on tractor at a Worcestershire demonstration in 1943. It had a recognisable two-wheel Trusty engine and transmission with the driving seat in a forward position and ran on Roadless Traction girder tracks.

Another ride-on prototype, the Trusty Motor Cycle Type Divisible Tractor, was described in the December 1944 edition of the Farm Implement & Machinery Review. The 6 hp three-wheel tractor had the same wheel layout as a motor cycle and side car. The purpose of the 'side car' wheel was to keep the tractor upright when it was used with a front-mounted cultivator, inter-row hoe or other tool, a plough offset from the rear wheel or a trailed implement. When more power was required the idea was to

2.68. This prototype ride-on Trusty tractor was demonstrated in 1944.

2.69. A rare Mk I Trusty Steed with a JAP 55 twin-cylinder engine.

remove the sidecar wheel and attach a second two-wheel Divisible Tractor alongside. If still more power was needed a third divisible Tractor unit could be added in the same way.

A more conventional-looking prototype four-wheel tractor was demonstrated in 1947. The first production model of the Trusty Steed, so called because it could be ridden like a horse, cost £200 when it arrived on the market in 1948. The Mk I Steed was little more than a towing tractor with a maximum payload of about 1½ tons which, according to sales literature, was giving highly satisfactory service to county councils, speedway tracks, dog tracks and farms.

Most Mk I Steeds had a 6 hp Norton or a JAP four-stroke engine. A few were sold with a 14½ bhp Norton Big Four and at least two left the factory with a twin-cylinder JAP 55 engine. A centrifugal clutch which 'ensured a silky start at all times' engaged the drive to a single forward and safety reverse gearbox with a top speed of 8 mph. A joystick was used to control the separate drive to each rear wheel and this was used to power steer the tractor by slowing one wheel when making tight headland turns.

The completely redesigned Mk II Trusty Steed announced in 1950 cost £179, pneumatic tyres adding an extra £25. Unlike the Mk I it had mid- and rear-mounted toolbars for rowcrop work, which were raised and lowered with a spring-assisted hand lever. Tractors (London) Ltd described the new model as 'a thoroughbred built in the tradition of the trim and spirited four-legged animal that had served man down the centuries'.

The Mk II Steed had a front-mounted Norton or JAP engine rated at a maximum of 14½ bhp, a multi-plate clutch controlled with a foot pedal and a three forward and one reverse gearbox with top speeds of 1¾, 3 and 4½ mph and 1¼ mph in reverse. Transmission was by roller chain and spur reduction gear to the rear countershaft, while dog clutches in both wheel hubs were used independently with hand levers or together with a foot pedal. The Mk II Steed

2.70. The Mk II Trusty Steed. (Roger Smith)

had independent rear wheel brakes, a parking brake, power take-off and the wheel track was adjustable in 2 in steps from 28 to 44 in for rowcrop work.

A few Mk II Trusty Steeds had steel wheels or Roadless Traction half-tracks for use on marginal or hilly ground. There was a choice of a vertical or horizontal exhaust and an engine bonnet was available at extra cost. A Trusty Steed with a front-angled dozer blade was another variant of the Mk II tractor which, complete with implement lift and main drawbar, cost £322 17s 9d in the mid-1950s. As well as the usual toolbar accessories other implements, including a two-furrow plough, a mid-mounted cutter bar mower, a trailer and a forklift were made for the Mk II Trusty Steed.

Tractors (London) Ltd also made the 39 in wide Trusty Road Roller, two

2.71. The Mk II Steed on Roadless half tracks.

models of the Trusty dumper with a 5 hp JAP petrol or Petter diesel engine and the Trusty Mowmotor mower-pusher unit with hand-propelled side-wheel and roller mowers.

Tractors (London) Ltd ceased trading in 1978 when Mr Knifeton of Bentley Heath purchased the remaining stock of Trusty products and spare parts. However, it was not the end of the Trusty Steed. The Trusty Tractor Co was still based at Bentley Heath in 1995 when it introduced a new four-wheel drive Trusty Steed. Built in the Czech Republic, it had a 33½ hp water-cooled three-cylinder Lombardini diesel engine, a hydraulically operated single-plate clutch and an eight forward and reverse speed shuttle gearbox.

2.72. The new Trusty Steed from the Czech Republic was launched in 1995.

Other features included a live hydraulic system with a separate oil reservoir, category I rear three-point linkage, hydrostatic steering, hydraulic disc brakes, power take-off and a de-luxe safety cab with optional air conditioning. The new Trusty Steed with grassland tyres and safety cab cost £12,900, a front three-point linkage and power take-off adding £1,300 to the price.

UNI-HORSE

Lea Francis Cars of Coventry first made the Uni-Horse tractors with a metal bonnet, black steering wheel and narrow mudguards in 1961 but within a year the company had ceased trading. Edward Williams Engineers, which formed a new company called Uni-Horse Tractors Ltd in Smethwick but later based at Droitwich, manufactured Uni-Horse ride-on tractors until the early 1970s.

The original Lea Francis Uni-Horse had a BSA 420 cc engine with an electric starter and a three forward and one reverse gearbox. The specification also included a diff-lock, independent brakes and hydraulic linkage.

The Mk 2 Uni-Horse with the same BSA engine under a fibreglass bonnet was introduced in 1962. The tractor had a three forward and one reverse gearbox, a clutch control that disengaged the drive when the brakes were applied and a white steering wheel. Wide mudguards were added a year or so later.

The Uni-Horse Six Eight, also made in the mid-1960s, was powered by a 9 or 10 hp four-stroke MAG petrol engine. The specification included a multi-plate clutch, roller chain drive to a two forward and one reverse gearbox doubled up with a high-low ratio gearbox, diff-lock, rear wheel drum brakes and a power take-off.

The Mk 4 Uni-Horse, with a constant mesh four forward and two reverse gearbox, drum brakes and a

2.73. Lea Francis made the first Uni-Horse tractors in 1961 but Uni-Horse Tractors Ltd, a company owned by Edward Williams Engineers at Smethwick, built this one in 1963.

deep bucket seat, was introduced at the 1965 Royal Smithfield Show. It remained in production until 1970 when it was superseded by the Model 700 Uni-Horse, the model number relating to its 700 cc Reliant car engine. The new model also had a Reliant gearbox and hydraulic rear wheel brakes, with power transmitted to the rear wheels through a differential and spur gear final drive.

WHEEL HORSE

The American Wheel Horse ride-on tractor with a three forward and one reverse gearbox, automotive steering, rear wheel brakes and a range of twenty different attachments, was marketed by Garden Machinery Ltd of Slough in the late 1950s and early 1960s. The Wheel Horse, complete with power take-off, belt pulley and manual lift linkage, cost £180 in 1959.

Wheel Horse tractors including the 6 hp Ranger, 8 hp Commando and 12 hp Raider were made in Indiana in America and in Belgium. They were imported by G D Mountfield of Maidenhead in the late 1960s. By the mid-1970s a wider family of Wheel Horse tractors available on the UK market included the Lawn Ranger, the Commando, the Raider, the Charger and the GT14.

The Lawn Ranger with a 7 hp four-stroke Lawson engine, three forward gears and a combined clutch and brake had replaced the earlier Wheel Horse Ranger. The Commando 7 and 8 with 7 and 8 hp Kohler engines and three-speed transmission, and the Raider 10 and 12 with 10 and 12 hp Kohler power units and six-speed transmission, had electric starting and a combined clutch and brake arrangement.

The Kohler-engined 12 hp Charger and 14 hp G14 had a hydrostatic transmission and limited slip differential. In the early 1970s G D Mountfield sold the D series Wheel Horse along with its own M25, M30 and M36 garden tractors and range of

2.74. A late 1950s Wheel Horse garden tractor.

2.75. The 12 hp C-121 Wheel Horse was made in the early 1980s.

attachments. The 20 hp D Series Wheel Horse had a twin-cylinder Kohler engine with electric starting, hydrostatic transmission, rear power take-off, hydraulic linkage and independent disc brakes.

Nine C series machines were among a dozen or so 8 to 16 hp Wheel Horse tractors marketed by G D Mountfield in the early 1980s. Most of them had a Kohler engine but a twin-cylinder 16 hp Briggs & Stratton power unit was used for the C-161 Twin with a six-speed manual transmission and C-161 Twin Auto with hydrostatic drive. There was a choice of a recoil or electric starter and manual or hydraulic lift for the 8 to 14 hp models and more than twenty different attachments were available to keep the tractors busy throughout the year.

The early 1980s range also included the 20 hp D series Wheel Horse and the Briggs & Stratton-engined 8 hp A-81 ride-on with a three forward and one reverse gearbox. The 20 hp D series tractor specification included a twin-cylinder Kohler engine with electric starting, hydrostatic transmission, disc brakes, power take-off and hydraulic linkage.

Wheel Horse UK, based at Reading, sold an even wider range of machines in the late 1980s, including the 100 series ride-on mowers together with 200, 300 and 500 series garden tractors. The 200 series with Kawasaki engines included the 12½ hp 212 and the 17 hp 227 garden tractors with either a five forward and one reverse transaxle or a hydrostatic transaxle.

A 12 hp Kohler engine powered the 312-8. The 316-8 had a twin-cylinder Onan 16 hp engine. Both 300 series models were equipped with a six forward and two reverse transaxle. A single-lever hydrostatic transaxle was used on the 500 series 518-H Wheel

2.76. The Wheel Horse 518-H was imported by Wheel Horse UK in the late 1980s.

Horse which had an 18 hp twin-cylinder Onan engine.

A single-lever hydraulic implement lift was standard on the 518-H Wheel Horse but the mid-mounted mower deck, rotary cultivator and grader blade for the 300 series tractors were raised and lowered with a hand lever. A manual implement lift was also used for the 200 series mower deck.

All Wheel Horse tractors could be used with a snow blower and a drawbar was provided for towing a trailer, a leaf sweeper and other equipment. In the early 1990s Wheel Horse garden tractors were imported by Toro Wheel Horse UK of Ringwood in Hampshire, a division of Toro Commercial Products of St. Neots.

2.77. The Wild Midget self-propelled toolbar.

WILD MIDGET

The self-propelled Wild Midget toolbar exhibited at the 1945 Royal Agricultural Show was a combined Hosier, Bomford, McConnel design made by MB. Wild & Co of 50 Pall Mall, London. It was a very basic three-wheeled chassis with a single steel cleated driving wheel at the rear and steered with two steel castor wheels at the front. The front wheel track was adjustable from 4 ft 1 to 5 ft 11 in.

The rear wheel was chain driven by a rear-mounted 3 hp JAP petrol engine and a three forward and one reverse speed gearbox gave a top speed of 2½ mph. An ultra-low gear attachment with an extra-chain driven reduction of 8:1 fitted between the gearbox and final drive and engaged with a dog clutch to give the driver a choice of six gears. In the ultra-low gear range the Midget had forward speeds of 5½ , 9 and 15 feet per minute. The driver, who sat close to the ground on an adjustable hammock seat suspended above the toolbar, had an unobstructed view of the rows of plants being hoed or cultivated.

The Midget had a dual-steering system with a foot-operated rudder bar and a hand lever both linked to the front wheels. The underslung toolbar in two sections 3 ft 6 in wide was raised and lowered manually. The dual-control steering system allowed the driver to steer the Midget with one foot on the rudder bar while raising or lowering the toolbar with the hand levers.

A second hammock seat could be attached to the toolbar when a second person needed to ride on the machine for close handwork and, if necessary, extra seats could be added to accommodate up to four people on the machine. The makers claimed that it was possible to drive the Midget in low gear and hand weed or thin rows of plants while steering the machine with the foot-operated rudder bar.

The basic Wild Midget without any tools cost £181, a four-row hoe attachment was £39 and a six row unit cost £54 11s 0d. A spare pair of hoe blades cost £1 17s 0d. Other attachments for the Midget included cultivator tines, ridge hoes and seeder units.

2.78. Slater & England made the Series I Winget 42 in the early 1960s. (Roger Smith)

WINGET

Slater & England, a company better known for its concrete mixers, made Winget four-wheel garden tractors. Many of the 500 or so tractors built at Gloucester were exported. The specification included a 7¾ bhp air-cooled Lister SR1 direct-injection diesel engine which used 1½ pints of fuel in an hour, a three forward and one reverse speed Newage gearbox, rear transaxle, diff-lock, adjustable wheel track and power take-off. The Winget had a live hydraulic system with a Dowty pump belt-driven from the engine crankshaft.

Three models of the Winget 42 were introduced in 1966. The basic Series I tractor with a hand lift cost £438 10s 0d, the Series II with hydraulic linkage was £492 and the Series III with hydraulic linkage and a front-end loader cost £560. Various rear-mounted implements including a plough, cultivator and transport box were made for the Winget tractor which, according to a 1967 advertisement, was easy to handle and just as easy to buy. About 500 Winget tractors had been made by 1968 when Slater & England sold the production rights and remaining stock of parts to MSW Machinery of Wood Vale in London.

WOLSELEY

Wolseley Webb sold mini-tractors made in America by MTD and a range of four-wheel ride-on mowers in the 1970s and early 1980s. Wolseley ride-on mowers included the 348 - with an 8 hp Briggs & Stratton engine and a 30 in rotary mower deck - and the 8 hp 30 in cut Wolseley 308. The 348 had a four forward

2.79. Made in 1973 this Wolseley 16HS mini-tractor has a 16 hp Briggs & Stratton air-cooled engine.

engines and an automatic drive transmission.

The 10 hp tierra gold and white Wolseley 10 mini-tractor had a Briggs & Stratton petrol engine, differential and transaxle, three-point linkage and power take-off. The specification for the Wolseley 16 and 16HS mini-tractors included Briggs & Stratton 16 hp engines with electric starting, transaxle, differential, disc brakes, power take-off and hydraulic linkage. The Wolseley 16 had a manual gearbox with four forward speeds while the 16HS was equipped with an infinitely variable speed Peerless hydrostatic transaxle.

Attachments included a 42 in cut mid-mounted rotary mower deck, plough, bulldozer blade and a rear-mounted 36 in wide rotary tiller. Wolseley ride-on mowers and mini-tractors were discontinued when Qualcast acquired Wolseley Webb in 1984. E P Barrus of Bicester sold virtually identical MTD 990/16 and MTD 990/16HS mini-tractors in the mid-1970s.

and one reverse manual gearbox while the 308 was equipped with an automatic drive six-speed transmission. The 26 in cut Wolseley 265 and 268 ride-on mowers had 5 and 8 hp Briggs & Stratton

Chapter 3

Ploughs, Cultivators and Hoes

PLOUGHS AND CULTIVATORS

Spades and forks have been used to work the soil for centuries and over the years even these simple tools have received attention from more than one inventive mind. The revolutionary Terrex semi automatic garden spade, first made in France, was designed to take some of the backache out of digging a garden. The Terrex spade is used by applying foot pressure to a spring-loaded lever at the base of the spade handle. With the blade a full spit deep a tug on the handle releases the spring which flicks the blade forward to loosen and partly lift the soil which is then turned by hand in the normal way. The Terrex spade cost £3 19s 6d post free in 1951, it was £7.95 in 1973 and by 1995 a tenfold increase had put the price at £79.95. Terrex also made a semi automatic garden fork in the early 1950s. The Easi-Digger attachment for garden spades and forks, which cost £1 9s 9d, was a cheaper way to ease the backache of digging. It consisted of a foot-operated lever with a powerful spring, which was attached to a spade or fork handle. With the tines or blade at full depth, foot pressure on the lever made it much easier to lift and turn the spit.

The 1950s Jobber motorised fork made by Robinsons Developments of Winchester had the appearance of a two-wheeled petrol-engined pneumatic drill. It was used with various soil-working tools including forks, cultivators and a ridging body. In operation the Jobber, with a JAP or Villiers two-stroke engine, had a clutch-and-crank assembly which gave the digging fork and other attachments a vertical thrusting action. The engine unit could be removed from its wheels and used with an adapted 12 in cut Webb Wasp lawn mower.

Revolutionary !!

THE TERREX SPADE

Send now for this NEW SEMI-AUTOMATIC Garden Implement and reduce your work by half.

UNCONDITIONALLY GUARANTEED

Simple to use—Easy to handle, will give years of effortless digging on all types of soil.

Price **79/6** Post free

3.1. *The Terrex semi-automatic spade was first made in the early 1900s.*

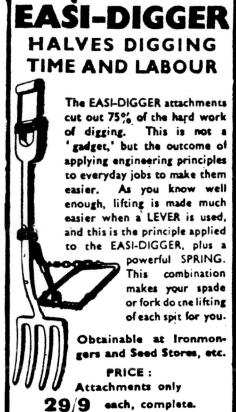

—THE—

EASI-DIGGER

HALVES DIGGING TIME AND LABOUR

The EASI-DIGGER attachments cut out 75% of the hard work of digging. This is not a 'gadget,' but the outcome of applying engineering principles to everyday jobs to make them easier. As you know well enough, lifting is made much easier when a LEVER is used, and this is the principle applied to the EASI-DIGGER, plus a powerful SPRING. This combination makes your spade or fork do the lifting of each spit for you.

Obtainable at Ironmongers and Seed Stores, etc.

PRICE :
Attachments only
29/9 each, complete.

3.2. *The Easi-Digger was claimed to take the hard work out of digging.*

3.3. A small plough, cultivating tines and hoe blades were included in the price of the single- and two-wheel Planet Junior push hoes.

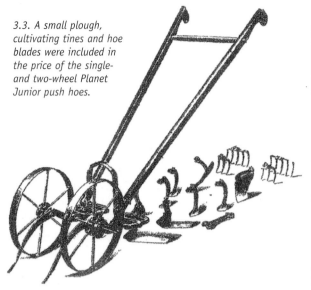

3.4. The Mayfield plough turned a furrow 6–7 in deep.

Cultivator tines and miniature plough bodies made for the Planet Junior and similar push hoes were quite effective on prepared soil but engine power made life easier when the ground was hard and dry. Small ploughs and cultivator tines were often included in the price of motorised hoes and some of these hoes could also be used with tools made for the Planet Junior push hoe but again they were more effective in loose soil.

The power required to break undisturbed soil with a plough or cultivator was provided by various types of two wheel garden tractor, ride-on tractor and small crawler. Most ploughs had a full mouldboard but the open-slatted mouldboard, sometimes used on farm ploughs, was an option in 1947 for the Greyhound plough made for the two-wheel Trusty tractor. Gravely made a rotary plough for the Model D and Model L tractors. The drive was protected by a slip clutch and the rotor, which turned at 800 rpm, ploughed a furrow 12 in wide and, depending on conditions, up to 7 in deep.

Some two wheeled garden tractors were single-purpose rotary cultivators but most of them were also used with a toolbar. The more powerful two wheeled tractors, including the British Anzani Iron Horse, BMB Plow Mate, Rowtrac and Trusty, sometimes advertised as an alternative to a horse, were used with a single-furrow plough, a cultivator, a ridging body, hoes and various types of harrow. A heavy counterbalance weight in front of the engine was recommended when using the single-furrow Oliver or Ransomes plough made for the Rowtrac 5 tractor.

A reversible plough for the Iron Horse was

3.5. The Barford Atom 15 and plough. (Roger Smith)

3.6. The Gravely rotary plough for the Model L tractor was introduced in 1938. (Roger Smith)

advertised as the finest of its type when it was introduced in 1950. British Anzani explained that although it had not pushed its sale it considered that all Iron Horse tractor owners should be made aware of the new plough before the waiting list for it piled up.

The Allen Motor Scythe was an unlikely power source for ploughing but a small front-mounted mouldboard plough (picture 3.8) in place of the cutter bar was already being used in the late 1940s. It was not the easiest of implements to handle and so the Allen plough was supplied with a front weight to help prevent it riding out of work and a vertical extension for the handlebars allowed the user to walk in a relatively upright position. An improved version introduced in 1956 with a heavier mouldboard eliminated the need for a front weight.

Walking behind a two wheeled tractor and single-furrow plough all day was hard and exhausting work but sitting on a BMB President, Byron, Garner, Gunsmith, OTA or other ride-on tractor was much less tiring and much more productive. Optional hydraulic linkage was available for the BMB President but a hand lever was used to raise

3.7. The British Anzani Iron horse and plough.

3.8. Ploughing with an Allen Motor Scythe with extended handlebars. (Roger Smith)

and lower a single-furrow mounted plough and the toolbar equipment used on most three- and four-wheel tractors.

The price of many small ride-on tractors was not very competitive when compared with the Ferguson TE 20 and its refined hydraulic system. For this

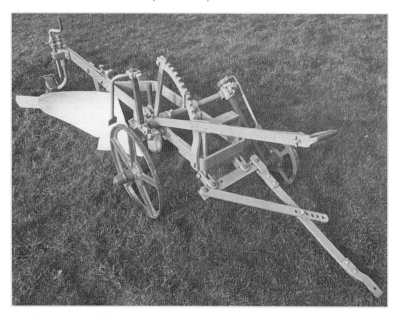

3.9. The Ransomes TS25 was made for MG crawler tractors.

reason many smallholders and market gardeners were able to justify the cost of a little grey Fergie which brought a newfound simplicity to ploughing, cultivating, discing, rolling and other fieldwork.

Ransomes, Sims & Jefferies of Ipswich made miniature versions of its trailed ploughs and cultivators for the MG crawler tractor. The TS 25 two-furrow trailed plough for the MG 2, introduced in 1935, could be converted to a single-furrow model and, depending on conditions, it turned a furrow between 8 and 10 in deep. The TS 30 single-furrow and TS 31 two-furrow hand-lift ploughs came next and both had a small screw handle on the rear wheel to keep the plough level from front to rear.

The self-lift single-furrow TS 42 plough launched in 1945 with the depth and levelling screw handles within easy reach of the driving seat was a great improvement on the earlier model. The TS 42A with low-level screw handles for ploughing under low branches in fruit

3.10. Tractors (London) Ltd. made single furrow ploughs for the Mk II Trusty Steed. (Roger Smith)

orchards was added in 1946 and the later TS 42B was a strengthened version of the original TS 42. The right handed TS 65 single-furrow mounted plough was introduced in 1951 for the MG 6, and MG 40 hydraulic linkage. Using the TS 66 with a left handed body made it possible to do reversible ploughing with the MG by hitching the two ploughs side by side on the linkage.

Compact tractors made in Japan were popular in Great Britain by the late 1970s and both Japanese and British implement manufacturers made ploughs and cultivators for the new generation of light tractors. Reversible ploughing was an almost universal practice in British agriculture by this time and the availability of reversible ploughs for compact tractors enabled market gardeners and smallholders to follow this trend.

Ransomes and other implement manufacturers made special-purpose ploughs for harvesting sugar beet, potatoes and root vegetables. Some smallholders endured the back-breaking work of harvesting potatoes with a wide-tined fork while others lifted the crop with a mouldboard plough. Both methods left many potatoes buried in the soil but the potato plough with prongs behind a small

3.11. Reversible ploughing with a Trusty two-wheel tractor. (Roger Smith)

ridger-shaped body did a better job by leaving more of the crop on the surface for hand picking. Potato ploughs were made for several garden tractors including the Iron Horse, Clifford and Ransomes MG crawler.

Potato spinners and elevator diggers were in use on many farms in the 1940s and 1950s when Clifford Aero & Auto made a potato spinner for its rotary cultivators. A wide share loosened the ridge and the spinner rotor, driven by the rotary cultivator power shaft, moved soil and potatoes sideways on to fresh ground to make hand picking potatoes a little easier than it was behind a potato plough.

Mechanical spading machines, which mimic the action of hand digging, have been used by some market gardeners as an alternative to ploughing. The Rotaspa spading machine was developed in Holland by Vicon in the early 1960s and the horticultural model had three sets of three mechanical spades with a working width of 42 in and a maximum digging depth of 12 in. A wider 7 ft spading machine was made for the agricultural

3.12. Reversible ploughing with a 12½ hp Kubota B 6000.

3.13. Ploughing with the Newman WD2 tractor. (Roger Smith)

3.14. The Vicon Rotaspa spading machine.

market. The power take off driven blades rotated around a horizontal rotor shaft and an arrangement of gears inside this shaft caused each blade to cut, lift and invert a block of soil before returning it to the ground.

3.15. Ransomes made a potato-lifting plough for the MG crawler.

Sales literature explained that the soil was left spaded in exactly the same way as hand digging and because the blades turned quite slowly soil smear and panning were prevented. The 42 in Rotaspa required at least 13 hp at the power take off and could dig about three-eights of an acre per hour at an average speed of 1 mph.

Pedestrian-controlled rotary cultivators described earlier have been used since the 1930s by gardeners and commercial growers to break up soil, prepare seedbeds and cut up weeds and crop residues. Some of them were single-purpose rotary cultivators but most two-wheel garden tractors had a rotary cultivator which could be interchanged with a plough or a toolbar for rowcrop work.

A new generation of three-point linkage rotary cultivators for compact tractors appeared in the mid-1970s. They were similar to the Howard Rotavator

3.16. The Clifford potato spinner.

3.17.
The Howard
Bullfinch
Rotavator.

3.18. The Mansley rotary cultivator was approved for use with the Ransomes MG crawler.

made twenty years earlier for the Ferguson 20 and other 20 to 30 hp tractors. Narrow-mounted rotary cultivators were usually off set on the tractor and removed one of the marks left by the rear wheel on the previous run.

Harrows, cultivators and other tined implements date back to the era of the horse. Various sizes of these implements were made for the Planet Junior and other push hoes, two- and four-wheeled garden tractors and market garden crawlers. A simple toolbar with a pair of wheels was either included in the price or supplied as an optional extra for most two wheel garden tractors in the 1940s and 1950s. The range of tools varied from a few cultivator tines and hoe blades to lengthy lists of tillage equipment made for the British Anzani Iron Horse, BMB Plow Mate, Trusty and other garden tractors.

An extensive range of cultivation equipment including tines and hoe blades was made for the front- and rear-mounted Ransomes MG crawler toolbars. The rear-mounted C29 and C71 toolbars were raised and lowered with the hand lever but the optional MG6 or MG40

3.19. Kubota made rotary cultivators for its compact tractors.

3.20. The Barford Atom and cultivator.

3.21. Ransomes MG 2 and cultivator.

3.22. Disc harrowing with a Trusty garden tractor. (Roger Smith)

hydraulic linkage was needed for the front-mounted C70 and rear-mounted C67 toolbars. A 3 ft wide set of Ransomes HR4 trailed disc harrows for the MG had 16 in diameter discs angled from the tractor seat. Tillage implements approved for use with MG included land rolls built by Hunts of Earls Colne and Gibbs of Bedfont, a steerage hoe made by G J Garrett & Sons of Dartford and the Mansley rotary cultivator from Gregson & Monk of Preston.

3.23. A roll was one of many attachments made for Clifford rotary cultivators.

3.24. The Massey Ferguson 1030 and cultivator.

HOES

In the early nineteenth century sowing seeds with a drill opened up the opportunity to hoe between the rows with a horse-drawn or hand hoe. At one time there were almost as many types of hand hoe with either long or short handles as there were county dialects. Examples include the swan neck or Stalham hoe, which took its name from a Norfolk village, the short-handled Lincolnshire hoe, the Dutch push hoe and the Millars' Safe-Speedy hand hoe. Made by Millars' Machinery in Bishops Stortford, the square-framed Safe-Speedy hoe was claimed to be an ideal tool for cleaning round and singling rowcrops. With a 1¼ in diameter handle the hoe cost 6s 9d in 1944.

One- and two-wheel push hoes appeared in the 1930s, the most famous being the Planet Junior from America which was sold in its thousands throughout the world. At least a dozen different models of the wooden-handled Planet Junior push hoe were listed in a 1939 garden equipment catalogue. The single-wheel No 8 with a pair of 6 in hoes cost £1 10s 6d.

Some single-wheel Planet Juniors could, with a special bracket and a second wheel, be converted to a two wheel hoe. The two-wheel Planet No 25 was more expensive at £5 18s 9d but hoe blades, cultivator tines, a small plough and a seeder unit were included in the price. The single-wheel hoe was pushed along between the rows with a pair of L hoe blades set as close as possible to the plants. Two wheeled hoes were used with the wheels either straddling a single row or with both wheels between two rows of plants.

Other push hoes made in the 1940s and 1950s included the Jalo, M & G, Pulvo, New Colonial and Wrigley. J T Lowe of Wimborne in Dorset made Jalo single- and two wheel push hoes and Jiffy seeder units. A 1950 advertisement described the new hoe with a solid rubber tyre as the triumphant result of countless experiments with various materials under actual working conditions which at last had made it possible to get rid of the back breaking work of hand hoeing. Potential buyers were also informed that the streamlined shape of the unit gave perfect control and even the most troublesome weeds were

3.25. Millars' Safe-Speedy hand hoe.

3.26. The Wrigley push hoe.

3.27. The M & G two-wheel hoe.

severed instantly! Supplied direct from the manufacturers, the Jalo hoe complete with a pair of hoe blades cost £3 7s 0d carriage paid. Attachments for cultivating, drilling, raking and ploughing in cultivated ground were available at extra cost.

J T Lowe, which adopted Jalo as the company name in 1951, was still making push hoes in the mid-1960s. The Jalo Gardener with a 'U' shaped tubular handle, widely advertised during the late 1950s, was claimed to hoe up to twelve times faster than it was possible by

hand. Complete with a pair of hoes and small plough it cost 5 guineas in 1959. Hire purchase terms were available with a down payment of £1 6s 0d and four instalments of the same amount. The Jalo Gardener could be used to ridge, drill, cultivate and rake while the two-wheel model could be converted to a wheelbarrow or fitted with a Sheen Flame Gun to burn off weeds.

Accessories for the M & G single- and two wheel push hoes made by Murwood Agricultural at Grosvenor Gardens, London SW1 and at Aldridge in Staffordshire included cultivator tines, rake, plough and a small ridging body. A 1950 Murwood sales leaflet included the single-row M & G push hoe with four cultivator tines and two 8 in hoe blades for £3 carriage paid; a seeder unit cost £2 5s 0d and a kit to convert the hoe to a wheelbarrow was £2. The two-wheel version of the M & G push hoe with hoe blades, cultivator tines and leaf guards cost £5.

Introduced in the late 1950s, the hand-propelled Pulvo garden hoe and cultivator with a patent vibratory action consisted of an open roller with toothed bars and a set of cultivating tines or hoe blades on a small tool frame. An advertisement

SAY 'GOODBYE' TO WEEDS —THE MODERN WAY!

JALO Hoes are designed for every GARDEN, ALLOTMENT, SMALLHOLDING, NURSERY & FARM and are the result of countless experiments under actual working conditions. Not until you have actually tried the JALO will you really believe how much *Easier, Quicker* you can remove weeds. Whether hoeing, cultivating, raking or ploughing with your JALO Hoe, you'll be delighted with its perfect performance.
PRICES : Single-wheel Model, 70/-; Twin-wheel Model, 105/-. Jalo Hoes are only obtainable direct from the Makers.

3.28. The Jalo push hoe.

explained that the Pulvo hoe pulverised the soil to stir up weeds or make a fine tilth faster and more efficiently than previously thought possible. It was also pointed out that it had nothing to go wrong, there were no extras to buy and there were no running costs. The Pulvo hoe cost 5 guineas but easy terms were available with a deposit of 12s 9d and eight fortnightly payments of the same amount.

Geo Bignell at Sutton Coldfield made the single wheel New Colonial push hoe and cultivator in the early 1950s.

The Wrigley push hoe manufactured by Wessex Industries at Poole from the late 1940s could be used to hoe, rake, plough and cultivate. The single-wheel Wrigley cost £4 5s 0d in 1959 and the two-wheel model cost an extra £2. An advertisement described the Wrigley hoe as undoubtedly the most practical implement obtainable at such a remarkably low price.

Industrial and Agricultural Developments Ltd of

Great Malvern made the Ro-Lo cultivator in the early 1950s. Publicity material explained that it bridged the gap between manpower and horsepower and offered a new era of ease and prosperity to all engaged in working the soil for profit or pleasure. The Ro Lo had a short toolbar for hoe blades and other accessories carried on a smooth roller. The price, including a pair of hoe blades, was £3 10s 0d, with cultivator tines and a ridging body available at extra cost.

The late 1940s Mitchell Colman Easy rotary push hoe was similar to the Pulvo hoe with blades set at an angle across the side discs. The makers claimed that the Easy Hoe would do the work of six to ten expert hand hoers, killing all the weeds and at the same time breaking up the soil surface to leave a fine tilth. It could be used to hoe either one or two rows and could be changed in a minute from a one- to two-row hoe without using any tools.

3.29 The Pulvo hoe was said to pound the soil to a fine tilth.

3.30 The Ro-Lo push hoe cost £3 10s 0d in 1951.

Some crops grown by smallholders and market gardeners had to be singled and this was usually done with a long-handled hoe. Higher work rates could be achieved with a singling or gapping process and to this end Douglas Bomford and Eric Alley introduced a hand-operated mechanical gapping machine in 1928. Strapped around the operator's waist the gapping machine had a pair of swivelling blades connected to levers on the handlebars. The operator was required to walk backwards using the blades to remove the weeds and unwanted growing plants as he or she moved along the row. The hands were left free to operate the levers and withdraw the blades from the row to leave single plants to grow at the required spacing.

A later version with two or three units attached to a tractor toolbar was operated by men walking behind the machine. An improved version, known as the 'Rapid-O', was introduced by M B Wild of 50 Pall Mall, London in 1947. It had seats for the operators who were required to keep the gapping units on the row and use foot pedals to withdraw the hoe blades at intervals to leaves selected plants to grow.

Hamilton Motors of Edgware Road, London introduced the hand-propelled Douglas two-wheeled hoe and controlled gapper in 1946. It was reported that the new machine could work at an average speed of 30 yd per minute, equalling the speed of four expert hand hoers. Two side hoes cut a 4 in wide band on each side of the row while a third blade ran along the row of plants. When the machine was used as a gapper, a lever on the handlebars was used to move the centre blade to one side to leave a single plant to grow. When hoeing a previously thinned crop the lever was used to withdraw the blade when encountering a plant.

An alternative Douglas automatic rotary gapper had two side hoes and a spider wheel with four curved blades. It was claimed to make 150 gaps per minute under normal conditions, which was five times faster than hand gapping. The space between each plant could be set at 8, 10 or 12 inches with three different bevel gears in the drive from the wheels.

Mechanical gapping was also carried out with the self-propelled Howard Mini-Gapper (page 112). This early 1960s rear-engined ride-on machine, which

3.31. The Clifford A1 garden tractor with front-mounted hoes.

could also be used as an inter-row hoe, was driven at right angles across the rows to leave small groups of plants for hand singling.

Small self propelled motor hoes, popular in the late 1940s and early 1950s, provided an easier and more efficient way of killing weeds in rowcrops. Hoe blades, consisting of 'L' or side hoes to cut close up to the side of each row and 'A' hoes to cut the centre ground between the rows were among the attachments available for many garden tractor toolbars. Small motor hoes such as the British Anzani, Colwood Hornet, Ransomes Vibro

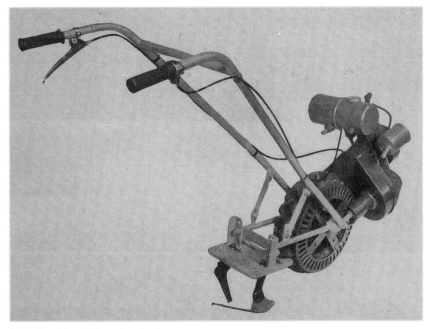

3.32. The Colwood Hornet motor hoe.

3.33. Planet Junior push hoe tools could be used with the British Anzani motor hoe.

Hoe, Commando, Coleby Minor, Mayfield, Teagle Digo and Wrigley were also used by market gardeners in the late 1940s and early 1950s.

Made by British Anzani Engineering Ltd at Hampton Hill in Middlesex and launched in the late 1940s, the British Anzani Motor hoe cost £39 10s 0d complete with hoe blades and grubbing tines. The 1 hp Anzani JAP four stroke engine ran all day on five pints of petrol and its forward speeds of 1 to 2½ mph were set with the throttle. The Anzani hoe had a centrifugal clutch with reduction gears and independent wheel clutches. It was only 8 in wide but the wheel track could be increased by adding spacers to the axles when hoeing in rows up to 13 in apart.

A sales leaflet pointed out that the Anzani Motor Hoe was not a tractor and, although it could also be used for cultivating, discing, seeding and light ploughing, it was only intended to do this work on properly prepared soils. It was claimed that its simplicity of control and its perfect balance gave expert results after very little practice and made work a pleasure.

Dashwood Engineering introduced the single-wheel Colwood Model A motor hoe in 1945. The Model A with a front bumper bar had a 1.2 hp JAP 2A engine to drive the single steel wheel through a three-speed gearbox. The Model B Colwood which superseded the Model A in 1949, had a two-speed Dashwood Albion

3.34. The Wrigley motor hoe. (Roger Smith)

gearbox and either a JAP 2A engine or a governor-controlled 1.2 hp Villiers Mk 10 four stroke power unit. Both engines were started with a hand lever. Hoe blades, cultivator tines, a grass mower, a hedge trimmer, a pump and a trailer were made for the Colwood Model B.

The Colwood Hornet motor hoe with a Villiers

3.35. The Wrigley had twin vee-belts to transmit drive from the JAP engine to the wheel.

Midget Mk II 98 cc two stroke engine appeared in 1951. The gear-driven land wheel was positioned centrally between the engine and the 9½ in wide toolbar. As well as the range of accessories made by Dashwood Engineering, Planet Junior push hoe tools could be used on the Hornet toolbar.

Wessex Industries at Poole in Dorset introduced the single-wheel Wrigley Wessex motor hoe in 1948 when it was advertised as the narrowest motor hoe on the British market. A 1 hp 98 cc four-stroke side-valve JAP engine mounted alongside the 25 in diameter single steel wheel powered the 10 in wide Wrigley motor hoe. Engine speed was automatically controlled with a hand-adjusted governor and a shaft drive from the engine was taken through a flexible coupling to a worm reduction box and twin vee-belts

3.36. The Coleby Minor motor hoe.

to a large diameter pulley on the wheel. A twist grip clutch control connected to a spring-loaded jockey-pulley was used to tension the drive belts and give a smooth take-off. A power take-off shaft was provided at the side of the gearbox and the tubular steel frame was equipped with a hinged parking stand.

Complete with a pair of 6 in hoe blades, the Wrigley Motor hoe with its bright red frame and light green wheel cost £50 in 1951. Optional accessories included tools for ploughing, cultivating, ridging and earthing up potatoes. It was reported in the farming press that the new hoe was easy for juveniles to use and the narrowness of the Wrigley Motor Hoe would allow its use in closely spaced rowcrops.

The Coleby Minor motor hoe with a JAP 2A engine and centrifugal clutch to engage drive to the wheels was less than 8 in wide. It had a top speed of 1½ mph and ratchets provided independent drive to the wheels. The Mayfield Engineering Hoe & Mow motor hoe introduced in 1961 was another single-wheel machine, which as its name suggested, could be used for inter-row hoeing and cultivating and for cutting grass with an 18 in rotary mower. It had a 2 hp fan-cooled four-stroke engine started with a hand lever, a two-speed gearbox with top speeds of 1½ and 2½ mph and a heavy-duty motor cycle chain drive to the pneumatic-tyred wheel.

The Croft Mayfield and Allen Mayfield versions of the Hoe & Mow made until 1975 had either a 3½ hp Briggs & Stratton or 2½ hp Villiers engine. An improved Arun Hoe 'N' Mow introduced in 1978, also with a Briggs & Stratton engine, could be used with hoe blades and cultivator tines on an 18 in toolbar or with a 3 ft mower cutter bar.

3.37. The Barford Atom 15 tool plate could be used for hoeing and other rowcrop work.

In the late 1940s small garden tractors such as the Pegson Unitractor and the American-designed Gravely were also used for hoeing in rowcrops. The single-wheel Gravely motor cultivator with hoe blades on a toolbar in front of the engine could be used to hoe rowcrops at speeds of 3 mph. The Pegson Unitractor (page 68) with a Villiers engine mounted inside its steel-rimmed wheel also had a front toolbar.

Tractors (London) Ltd and Jalo combined their resources to produce the Jalo Trusty motor hoe in the early 1950s. It had a ⅓ hp engine with a chain drive to friction rollers held against the hoe wheels by the weight of the engine. Drive was engaged with a centrifugal clutch and a hand lever was used to lift the engine unit off the wheels so that the hoe could be hand-propelled in confined spaces. Publicity material at the time suggested there would be no more struggling or straining with out-of-date hand hoes and no more grunts and groans from back-breaking hand work as the Jalo Trusty hoe would easily and quickly remove weeds ten times faster than before.

The Trusty Weed Sweeper, a small rotary cultivator, and the Whirlwind combined rotary hoe and grass cutter both cost £30 ex works when they were introduced by Tractors (London) Ltd in 1957. The Weed Sweeper had a ⅓ hp engine with a vee belt drive to a 6 in wide rotary hoe. The same engine was used for the Trusty Whirlwind with transmission by vee belt to a rotor disc with a set of pins to remove weeds. The disc could be replaced with a cutter bar to deal with rough grass.

3.38. The Trusty Jalo motor hoe.

The J C Rotor Hoe (picture 3.39), made in the mid-1950s by A M Russell in Edinburgh, was an unusual hoe attachment for one and two wheeled garden tractors. It consisted of a horizontal weed-cutting blade below and behind a ground-contact driven rotor similar to a lawn mower cylinder with seven inclined blades. The Rotor Hoe was made in 8, 14 and 20 in widths and all three models could be extended by four inches to hoe 12, 18 and 24 in rows. The narrowest Rotor Hoe with three cutter blades cost £9 15s 0d in 1953. Sales information, which carried the slogan 'one year's seeds makes seven year's weeds' suggested the J C Rotor Hoe was particularly efficient in reasonably dry soil conditions and made weeding by normal hand hoeing obsolete.

The Commando power-driven hoe was described as the only powered draw-hoe capable of working in any soil. It cost £65 when introduced by Power Hoes of Ixworth Place, London SE3 in 1950 but within a year the price had risen to £75. (Picture 3.40).

The pneumatic-tyred wheels and draw-hoe blades were driven by a 1 hp four stroke side-valve engine which used a pint of petrol per hour. The throttle and clutch levers were attached to the adjustable handlebars, which could be offset in nine different

3.39. The J C Rotor Hoe.

positions. A single-plate clutch engaged the drive to the hoe blade crank mechanisms and a dog clutch was used to engage drive to the wheels. Hoeing width was instantly adjustable with a sideways movement of the right-hand handlebar. The Commando could be used with 7, 9 or 11 in wide blades to hoe in rows between 9 in and 22 in apart and up to 3 in deep in a single pass.

An advertisement introducing the Commando power hoe claimed that it was a big advance in powered hoeing and that it could do everything at least five times faster than an expert with a draw hoe. The Commando was said to go anywhere and tackle anything. It could work against a wall or hedge,

incorporate mulches or fertiliser, earth up plants or scrape soil away from the base of onions and similar crops.

Mid-1950s sales literature for the self propelled Ransomes Vibro Hoe explained that the unique principle of its reciprocating hoe blades made it the most efficient hoe and cultivator ever offered to smallholders and horticulturists. Furthermore, it would banish the tiresome back-aching drudgery associated with manual hoeing. The prototype version of the Vibro Hoe had two wheels but to keep it as narrow as possible production machines had a single wheel with a solid rubber tyre.

Initially made in 1954 by farm baler manufacturer

D Lorant Ltd of Watford (which was acquired by Ransomes in 1951), the Vibro Hoe had a 98 cc two-stroke Villiers engine with a centrifugal clutch to engage drive to the land wheel and hoeing mechanism. The throttle had to be fully open when starting the engine so a dog clutch was provided to disengage the drive to the land wheel and hoeing mechanism before using the starter rope. A system of cranks was used to move the blades alternately forwards and backwards, giving them a walking action as the machine progressed along the row.

Separate drive trains were used for the hoe blades and the land wheel on the Ransomes Vibro Hoe. At least four variations of the double crank drive arrangement for the hoe blades were used during the machine's four year production run. The direct drive from the engine to the land wheel was through a chain reduction drive and two pairs of 5:1 reduction gears.

The number of hoeing strokes per minute was directly proportional to the Vibro Hoe's forward speed. Short lengths of bicycle chain were attached to the back of the blades to stir the soil and expose the weed roots to the drying sun and wind. The hoe blades were 6 in wide but rows

3.40. The Commando power-driven hoe.

3.41. A double crankshaft arrangement was used for the walking hoe blades on the MK.3 Ransomes Vibro Hoe.

3.42. The first Ransomes Vibro Hoes were made by D Lorant Ltd at Watford.

3.43. Inter-row hoeing with the Bean rowcrop tractor.

up to 18 in apart could be hoed with extension bars on the blades. The Vibro Hoe's vibrating motion was also an advantage when using cultivator tines.

The Vibro Hoe was mechanically suspect and all machines sold were recalled on a couple of occasions for modifications at the factory. The dog clutch used to disengage the drive before starting the engine had a tendency to seize and when it was dismantled it was not unusual for springs and ball bearings to fly in all directions. Production was transferred from Watford to Ipswich in 1956 and the last Vibro Hoes were made at the Orwell Works in 1958.

A great deal of walking was necessary when using a push hoe or hoeing rowcrops with a two-wheel garden tractor. By the late 1940s some large acreage growers were hoeing four or six rows at a time with a Ransomes MG crawler or a self-propelled toolbar. The first Bean self-propelled toolbars (page 95) were sold in 1946 and the Wild Midget toolbar (page 146) was shown at the 1947 Royal Show. Numerous tractor-mounted steerage hoes had already brought welcome relief to tired legs long before the David Brown 2D tool carrier appeared in 1955.

3.44. The 1962 range of Sheen flame guns.

Burning off vegetation with a flame gun was an alternative method of killing weeds and this system of weed control had much in its favour when dealing with perennial weeds. Several companies, including Bering Engineering of Camberley, McAllan Appliances of London and Sheen of Nottingham, sold flame guns in the 1950s. Sheen introduced its flame guns early in the 1950s and within ten years the company was making fourteen different models, some of which were still being made in 1975.

Sheen flame guns ranged from a small hand-held torch to large flame guns for use with small garden tractors or push hoes. A 1962 Sheen advertisement admitted that a flame gun would not kill perennial weeds in one operation but explained that its two stage system gave fully·effective weed control in a fraction of the time taken by conventional methods.

The Sheen X300, a typical hand-held flame gun, had a 1 gallon fuel tank, a hand-operated pump to pressurise the container, a pressure gauge, a control valve and a burner which produced a flame of approximately 2,0000F. A hood was used to restrict the spread of flame in confined spaces. A wheeled version of the X300 gun had a hinged hood and was recommended for inter row weeding and for burning off weeds on paths and crazy paving.

The Sheen Flamewand, resembling an overgrown bicycle pump with its tank holding enough fuel for about thirty minutes' work, produced a torch like flame suitable for weeding in very confined spaces.

3.45. The Bering flame gun.

Chapter 4

Sowing, Spreading and Spraying

Mechanical aids for sowing seed and spreading fertiliser were being used on some smallholdings and market gardens in the 1930s and 1940s but the home gardener did this work by hand. A variety of spraying and dusting machines for the distribution of liquid and powdered chemicals was also available at this time but their designers appear to have taken little notice of the health risks involved for those using this equipment.

SEED DRILLS

Small hand-held seed sowers were a useful aid to help home gardeners plant their seed sparingly and reduce the time taken at a later stage to thin the crop. Broadcasting grass and clover seed was often done by hand but some was sown with a seed fiddle. Sugar beet and all sorts of vegetable seeds from lettuce to peas were still sown continuously in rows, usually with a hand-propelled brush feed drill. By the late 1940s however, some growers had mechanised this work by using two or more seeder units on a garden tractor toolbar.

Brush feed drills, which date back to the early 1900s, were still in common use in the late 1940s. The Feedex brush feed drill had a land wheel-driven rotary brush on a shaft in the bottom of the seed

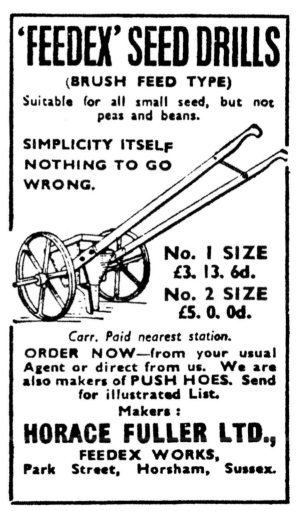

'FEEDEX' SEED DRILLS

(BRUSH FEED TYPE)

Suitable for all small seed, but not peas and beans.

SIMPLICITY ITSELF NOTHING TO GO WRONG.

No. I SIZE
£3. 13. 6d.
No. 2 SIZE
£5. 0. 0d.

Carr. Paid nearest station.

ORDER NOW—from your usual Agent or direct from us. We are also makers of PUSH HOES. Send for illustrated List.

Makers :

HORACE FULLER LTD.,

FEEDEX WORKS,
Park Street, Horsham, Sussex.

IT'S SOW EASY

Make every crop a competition winner by sowing with a "Shakit."

- Even and rapid sowing.
- Adjustable for different sized seeds.
- Nothing to wear out.
- Reduces need for thinning, thus promoting sturdier growth.
- Indispensable to the serious gardener.

W. E. BEVAN & CO.,

2/6 PLUS 3ᵈ POSTAGE

4.1. The Shakit Seed sower was a useful gardening aid.

4.2. The Feedex seed drill was made in the early 1950s.

hopper and a simple coulter to make a shallow furrow. The brush swept seed through an adjustable-sized hole in a disc near the base of the hopper where it dropped into a short tube connected to the coulter. The metal seed disc with various diameter holes was rotated to give the required size hole for different seeds and to vary the seed rate. A hand lever was provided to close the hole at the end of a row to cut off the flow of seed.

Seeder units were among the toolbar attachments for Barford, Coleby, Trusty and other two-wheel tractors. The Coleby seed drill with patent sweep discs on a wheel-driven shaft in the hopper gave a more positive delivery of seed and was an improvement on the earlier and rather inaccurate brush feed mechanism. Seed spacing depended on the number of discs on the shaft and so the feed mechanism could be set to space seed at 10, 15 or even 30 in apart, saving a considerable amount of hand work when singling the crop. An advertisement in 1950 for the two-row Barford Atom seeder, which cost £16 0s 0d ex works, explained that it would cut a furrow, sow various sizes of seed and cover them with soil in one operation.

The drill unit for the American Planet Junior push hoe, used to sow seed continuously in rows or in small groups, had a set of star-shaped land wheel-driven cams with each cam giving a different seed spacing. The cam opened a shutter in the bottom of the hopper to release a few seeds into a shallow furrow at intervals of 4, 6, 8 or 12 in. The release shutter was held open when drilling a continuous row seed.

A 1939 garden equipment catalogue noted that the Planet Junior drill was easy to push, sowed seed evenly and was suitable for practically all vegetable seeds and some flower seeds. Complete with a marker, the No 3 drill with a 3 quart capacity hopper cost £4 19s 9d and the No 5 with a 5 quart hopper was £5 11s 9d. The larger two-wheel No 25 combined hoe and seeder was recommended for the larger garden; when used as a hoe or cultivator the two wheels straddled the row of plants.

A later version of the Planet seeder, suitable for an even wider range of seeds, was either hand-propelled as a single row drill or as a multi row drill with between two and four units bolted on a garden tractor toolbar. The land wheel-driven seeding mechanism consisted of a wavy agitator disc, which directed seed through a round hole in a circular seed plate at the bottom of the seed hopper. The planet seeder unit was supplied with a set of four seed plates, each with a series of different-sized holes; one set was sufficient to sow any size of seed from clover to beans. The required seed plate was clipped in position under the hopper and rotated to align the chosen size of hole with the outlet point above the coulter.

Humberside Agricultural Products, which made the Bean self propelled toolbar, used the Planet Junior feed mechanism for the new Bean hand-propelled seed drill introduced in 1947. Designed for market garden use, the Bean drill

4.3. The star-shaped cams below the Planet Junior seeder handlebars were used to vary the seed rate.

4.4. The Bean hand drill with spare seed discs clipped to the handlebars cost £9 7s 6d in 1952.

consisted of handlebars and frame with a drive wheel, rear press wheel, seed hopper and four seed discs.

The seeding mechanism was similar to that used for the later type of Planet seeder with a rotary agitator ejecting seeds through one of the forty-six holes with sizes ranging from ³⁄₃₂ in to 1 in in diameter. The choice of hole depended on the type of seed and required seeding rate. The hand drill hopper held four quarts of seed, a shut-off lever was provided on the handlebars and a marker scratched a line as a guide for the next row. As pea and bean seeds were sold in pint and quart packets these units were given in sales literature and instruction books.

The Whitwood seed drill first made in 1948 by Dashwood Engineering could be used with the Colwood garden tractor or as a single-row hand-propelled drill. The land-wheel driven brush feed mechanism with a spacing disc was reputed to give an

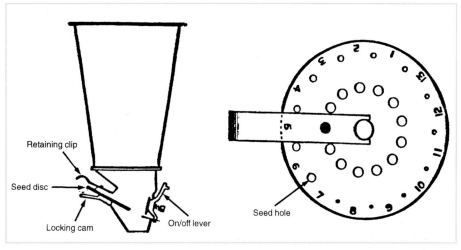

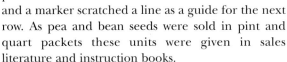

4.5. A rotary agitator in the Bean seeder hopper metered the seed through the selected hole in the seed disc.

even distribution of seed and accurate spacing in the row. The Whitwood seed drill with a larger hopper was also used to distribute fertiliser.

The 'One by One' multiple row seeder and hoe introduced in 1949 by Precision Seeders Ltd of West Wittering in Sussex was used as a hand-propelled drill to sow four rows at a time. As many as five seeder units could be attached to the smallest garden tractor or motor hoe toolbar. Adjustment was provided to

4.6. Bean seeder units on a Bean self-propelled toolbar.

4.7. Three Bean seeder units on a British Anzani Iron Horse toolbar.

4.9. The M & G seed drill hopper held seven pints of seed.

4.8. The Coleby seed drill was made in the 1950s for various models of garden tractor.

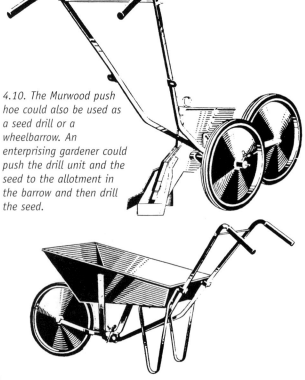

4.10. The Murwood push hoe could also be used as a seed drill or a wheelbarrow. An enterprising gardener could push the drill unit and the seed to the allotment in the barrow and then drill the seed.

space the independently mounted units at 3 in intervals and the mechanism could be set to deliver single seeds at any desired spacing to reduce hand thinning to a minimum. It only took a few minutes to change the seeder to an inter-row hoe. The wheeled drill frame and toolbar with one seeder and hoe unit cost £12 18s 6d, additional seeder units adding £8 10s 6d to the price.

A two-row seeder which cost £16 0s 0d was included in the 1950 list of attachments for the Barford Atom tractor. The seeder unit, which was only suitable for the two-wheeled Atom tractor, made two shallow furrows, planted the seed and covered them in one operation.

The seed hopper on the M & G drill unit made for the Murwood push hoe held one pint of seed and the agitator feed mechanism could sow any seed up to the size of a pea. The unit, complete with its own drive wheel, was attached to the Murwood push hoe frame and secured with a thumb screw in a matter of seconds. A Perspex panel in the hopper enabled the user to check seed flow to the coulter.

In 1947 Clifford Aero & Auto introduced a three-row seed drill on a toolbar for the Clifford rotary cultivator which, after the cultivating rotor was removed, could be attached to the tractor in a matter of minutes. Each seeder unit had a brush feed mechanism individually driven by the front wheel, while a coulter made a small furrow for the seed which was covered with soil and consolidated by a rear press wheel. The Clifford drill unit was suitable for any seed up to the size of a pea and, when drilling closely spaced rows, up to five units could be used on the toolbar.

Some farm machinery manufacturers who introduced toolbar-mounted precision seeder units for tractors in the early 1950s also made one-row hand-propelled seeders for market gardens. Each precision seeder unit had a front driving wheel, coulter, coverer and rear

press wheel suitable for vegetable crops and sugar beet. The Cuthbertson seeder unit was an early example but the Stanhay belt feed and the cell wheel mechanism used by Webb, Russell and other makers were more successful and have withstood the test of time.

The Cuthbertson unit drill, made at Biggar in Scotland in the late 1950s, had a small land wheel-driven auger revolving in a slotted bush in the bottom of the seed hopper. A rotating rubber agitator disc above the auger ensured a constant flow of seeds from the hopper through the spaces between the auger flights to the outlet point where they fell into a shallow furrow. Different-sized seed augers and slotted bushes were supplied with the seeder unit for sowing various types of seed at different rates. The single-row hand-propelled Cuthbertson drill cost £33 and individual units for tractor toolbars were £30.

Belt and cell wheel seeder units could be used in multiples on a rear-mounted toolbar or as a single-row hand-propelled drill for small-scale vegetable

4.11. Stanhay introduced the belt-feed precision seeder in 1953. The single-row hand-propelled drill was ideal for allotments and market gardeners could use two or more units on a garden tractor toolbar.

production. Market gardeners with large areas of vegetables used two or three of these seeder units on a two wheel garden tractor toolbar. Belt feed precision seeder units introduced by Stanhay at Ashford in Kent in 1953 are still being made in the twenty-first century. The Stanhay has a narrow rubber belt with holes at equal intervals to carry the seed from the hopper to an outlet point above the coulter. The size of the holes and the distance between them catered for different types of seed and their spacing in the row.

The cell wheel mechanism uses a similar principle but the seed is carried from the hopper in evenly spaced holes around the outer rim of the feed wheel to the outlet point above the coulter. Each seeder unit was supplied with a set of cell wheels for different types and spacing of the seed. The Russell Excel cell wheel hand-propelled seeder, which cost £80 in the late 1970s, had a two-pint capacity hopper and the aluminium cell wheel was driven by its rubber-tyred land wheel. Russell sales literature indicated that seed could be spaced at 1 or 1½ in apart or at any other spacing divisible equally into 72. Pelleted seed was in common use by the late 1970s and this contributed to the considerably improved accuracy of seed spacing with belt and cell wheel precision seeders.

The French SM frame drill made from aluminium

4.13. The Russell Excell hand drill was made in the mid-1970s.

consisted of a bank of four small diameter feed rotors driven by a square shaft from the land wheels. It was small enough to sow four closely spaced rows at a time in a cold frame. Different feed rotors were used to suit the type of seed and vary seed spacing in the rows.

FERTILISER SPREADERS

A great deal of fertiliser was spread by hand in the late 1940s and this method was not limited to allotments and market gardens. Where space allowed, a horse-drawn fertiliser distributor was used but many tons of fertiliser were spread by hand from a bucket often carried on a piece of rope slung around the shoulders. Still more handwork was required on holdings where it was the practice to mix two or more straight fertilisers with each containing one of the major plant nutrients. Compound fertilisers were available but many growers preferred to mix their own and save money.

Fisons, Sisis, Wolf Tools and others made small 12 to 36 in wide hand-propelled fertiliser spreaders in the 1950s and 1960s for the application of lawn sand and fertilisers. Many of them had a grooved or notched wooden feed roller, with a small land wheel at each end, in the bottom of a small triangular-section sheet steel hopper.

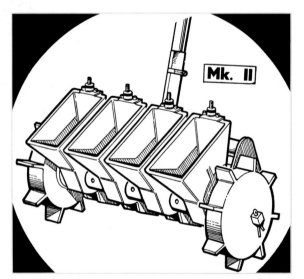

4.12. The French SM frame drill was used to sow four closely spaced rows at a time in a frame or small glasshouse.

The 12 in wide Fison spreader cost £2.68 in 1973 when a 24 in Wolf spreader was £19.95. Using different feed rollers varied application rate and some spreaders had an adjustable shutter to control the rate of flow from the hopper. Non-corrosive plastics were used for the hoppers and feed rollers in the late 1970s when the 12 in wide Fison spreader with a polypropylene hopper and a shut off lever cost £8.40. Although not a common practice, fertiliser was sometimes applied with a spreader attachment on a pedestrian-controlled garden tractor.

The Whitwood fertiliser spreader made for the Colwood garden tractor by Dashwood Engineering was a conversion of the Whitwood seed drill. A larger hopper was used when spreading fertiliser which could either be broadcast or placed by metal chutes between two rows or on both sides of a single row. The Whitwood could be supplied with handlebars and used as a hand-propelled spreader.

Broadcasters and full-width distributors, either towed or mounted on the hydraulic linkage of a compact tractor, were used to spread fertiliser on market gardens, golf courses and other large areas of grass in the 1950s. Some smallholders and market gardeners used a spinning disc broadcaster or a Vicon Varispreader which distributed the fertiliser with an oscillating spout.

4.14. A hand-propelled fertiliser spreader for lawns and vegetable gardens.

4.15. The Bean self-propelled toolbar with a fertiliser spreader.

The Farmaid Minor fertiliser distributor, made by New Harvest at Kidderminster for British Anzani, BMB Plowmate, Trusty and similar two-wheel tractors, was launched at the 1949 Royal Show. The Farmaid Minor was towed from the tractor drawbar and the outfit was steered from a seat on the hopper lid. It had a 2 to 3 cwt capacity hopper with a 4 ft spreading width and a land wheel-driven agitator spreading mechanism.

Full-width fertiliser distributors, mostly with plate and flicker or reciprocating plate feed mechanisms, were also used but they were difficult to clean and many suffered from the corrosive action of fertiliser.

4.16. The agitator feed mechanism in the Warrick fertiliser spreader was also suitable for broadcasting a wide range of seeds. The lightweight 7 ft wide hopper was tipped upside down for cleaning.

Less complicated feed mechanisms, usually with rotating agitators spaced at intervals in the hopper and driven by the land wheels, appeared in the mid-1950s. Fertiliser was agitated through adjustable-sized outlets, and deflector plates ensured even distribution across the width of the machine.

SPRAYERS

Gardeners and smallholders have used hand-operated and horse-drawn sprayers to apply pesticides, weed killers and other chemicals to their orchards, fields and gardens for the best part of a century. By the mid-1940s they were using hand-held sprayers and sprayers mounted on garden tractors for lime washing, glasshouse shading, pest control and killing weeds. Domestic gardeners of the day also gained benefit from applying dilute chemicals,

usually mixed in a bucket, with a hand-operated syringe or stirrup pump.

Allman, Cooper Pegler, A & G Cooper, Dorman, Drake & Fletcher, Evers & Wall, Four Oaks and W Weeks & Son were among a long list of companies making horticultural sprayers in the 1940s and 1950s. Their products ranged from hand-held syringes for the home garden to knapsack, barrow and garden tractor sprayers.

Hand-held syringes were sold by ironmongers and gardening shops in a variety of shapes and sizes. Most were made from non corrosive brass and the cheapest were sold for a few shillings. Syringes were filled by placing them in a container of chemical and withdrawing the plunger handle to suck liquid into the barrel. It was possible to spray almost non stop when using more expensive hand syringes with a

4.17. Hand syringes used to control pests in the garden and small greenhouses.

4.18. A selection of mid-1950s Solo garden sprayers.

4.19. This ASL compression sprayer cost £4.23 in 1973 but within five years inflation had more than doubled the price to £11.53.

suction hose placed in a container of diluted chemical.

The Four Oaks Streetly sprayer was a small hand-held pneumatic sprayer suitable for applying chemicals in glasshouses. An integral hand pump pressurised the three pint container making it possible to spray non-stop until the tank was empty. The Streetly cost £5 12s 0d in 1959, while the de luxe version with a thumb-operated control lever was an extra 9s 0d.

There was a wide choice of pneumatic and small hand-held sprayers in the 1970s for the domestic gardener. Compression sprayers with a one-gallon plastic tank cost between £5 and £10 and there was change from a £1 note for a small trigger-operated mist sprayer.

Mr E J Allman started a motor engineering business in a small shed in 1919 and within a few years he had moved to larger premises at Birdbrook near Chichester where he added agricultural engineering to his business interests. The first Allman sprayer did not appear until 1946 but since then the company has become a market leader with its range of knapsack, barrow and tractor spraying and dusting machinery. The Pestmaster Minor dusting machine and the Speedesi powder duster patented in 1947 were among the first Allman products.

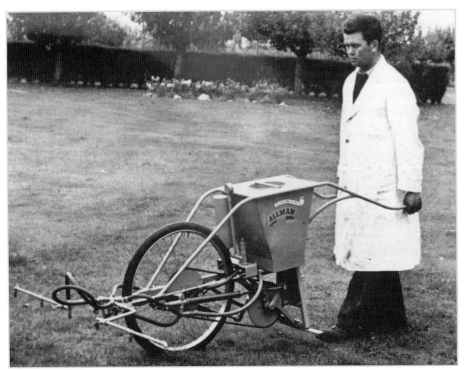

4.20. The makers claimed that a trial would soon convince potential buyers that the Allman Rapid sprayer was the most simple, efficient and easily manipulated sprayer available to the market gardener.

4.21. The Colwood motor hoe with an Allman sprayer.

The Rapid sprayer for weed control in rowcrops, seedbeds, nurseries and sports turf, was originally made by a Kent company in 1947 but Allman acquired the design in 1949. It had two brass pumps driven by a system of cranks from a bicycle wheel, a six-gallon galvanised steel tank, a foot-operated on/off control lever, and a pressure chamber was used to maintain a constant working pressure. Three nozzles on each half of the two-section spray bar were spaced at 12 in intervals and the bar could be set to a maximum height of 18 in above the ground. The Rapid was also used with the two sections in a vertical position for spraying fruit bushes. Application rates of 11 to 130 gallons per acre were possible by using different nozzles and varying walking speed.

A powder duster version of the Mk I Rapid with the fan driven by a belt from the land wheel was also made. The Mk II sprayer with a JAP four-stroke engine and 6 ft spray bar replaced the wheel-driven Rapid in the early 1960s. The JAP engine was used for a year or so but by 1963 it had been replaced with an American-built ½ hp Ohlsson & Rice two stroke engine. The Mk III Rapid with a new type of pump and a 6,000 rpm Villiers two stroke engine cost £78 17s 6d when it was announced in 1966.

Allman introduced sprayer units for the Colwood motor hoe and Landmaster rotary cultivator in 1959. The Colwood unit had a roller vane pump with a maximum working pressure of 100 psi. The pump was vee belt driven through a 5:1 speed reduction from a pulley on the engine crankshaft. The specification included an 8 gallon galvanised tank, pressure regulator, pressure gauge and a two-way on/off tap. The two-part spray bar could be used for vertical and horizontal applications, and hand lances were provided for spraying trees, hedgerows, buildings, etc.

The Allman sprayer for Landmaster Gamecock and Kestrel rotary cultivators used the same roller vane pump attached to the rotary cultivator's drive shaft. A 20 gallon tank, close-coupled to the tractor, was carried on small castor wheels at the rear. Controls were similar to those on the Colwood unit; hand lances were available and the spray bar could be set for vertical or horizontal spraying.

The Allman Gardenspray was originally manufactured in 1959 for the Landmaster range of Gardenmaster garden cultivators but within a year or so it was being made for other makes of garden cultivator with a front rotor. The Gardenspray had a 2 gallon plastic tank, brass piston pump, pre set relief valve and a hand lance with an on/off tap. The pump, which delivered up to 30 gallons per hour at a pressure of 40 psi, was driven from the cultivator rotor shaft.

The hand lance had three different nozzles with spray patterns from a fine mist to a pencil jet used to apply winter wash to fruit trees. Return flow of chemical from the relief valve was used to agitate the tank contents and keep them well mixed. The moving parts ran in oil-impregnated bushes and no tools were required when fitting the sprayer attachment to the Gardenmaster.

The Allman Polypak Kestrel knapsack sprayer introduced in 1961 was suitable for a wide range of chemicals, whitewash, creosote, etc. After initial pressurising with the diaphragm pump, a few strokes with the pump handle were enough to maintain the compression cylinder at its 65 psi working pressure.

The Midget Wet sprayer for the application of most types of insecticide, fungicide, herbicide and

4.22. The spray bar on the Allman Midget wet sprayer could be used either horizontally or vertically and two hand lances were provided for spraying fruit bushes and trees.

4.23. The Allman Autospray was approved for use with Auto Culto garden tractors.

4.24. The Byron Landmaster high-pressure spraying unit..

disinfectant was added to the Allman range in 1963. A 1½ hp Villiers or 1¾ hp JAP four-stroke engine, which could be adapted to run on tvo at a small extra cost, was used to drive the pump. The first Midget sprayers had an Allman gear pump made from bronze but subsequent machines had a more wear-resistant roller vane pump.

The Midget barrow sprayer had a 6 gallon tank on a two wheel trolley and three low-volume nozzles on a 4 ft 6 in spray bar. The spray bar could be extended to 8 ft 4 in for sports fields and other large areas of land. Hydraulic agitation was used to keep the tank contents well mixed and operating pressure was up to 140 psi depending on the size of the nozzles. Allman may have looked into the future with a crystal ball but it was more likely that the Midget wet sprayer would be

4.25. The Allman Rapid Mk V barrow sprayer, which cost £397 in the mid-1970s, had a twin-diaphragm pump driven by a four-stroke petrol engine.

supplied with a pressure gauge calibrated in psi or in bars to meet the needs of its Continental customers.

The Allman Autospray and Merryspray were introduced in 1959. The Autospray was approved for use with the Auto-Culto 65, Mk IX Autogardener and the Midgi-Culto while the Merryspray was recommended for the Wolseley Merry Tiller. Both sprayers had a roller vane pump with a maximum output of 2½ gallons a minute vee-belt driven from a pulley on the engine crankshaft.

Chemical from the 5½ gallon tank mounted on the handlebars was applied by three nozzles on each half of the two-section spray bar at a rate of 11 to 130 gallons per acre. The spray bar covered 6 ft when working in rowcrops and the two spray bar sections could be used in a vertical position for spraying fruit bushes. A hand lance and pressure washing gun increased the versatility of both machines.

The Minispray knapsack sprayer was added to the Allman range in 1966. The lightweight sprayer with a transparent 2½ gallon plastic tank, piston pump and

plastic hand lance weighed a mere 6 lb when empty. Minispray accessories included a hand-held dribble bar applicator, an Arboguard attachment for ring spraying around young trees and an Expando telescopic dribble bar with adjustable crop shields to cater for different row widths.

The improved Allman Arbogard Mk BI launched in 1969 was, like the earlier version, used with a knapsack sprayer to ring-weed young trees with Gramoxone. The Arbogard – an ICI trademark – had a single flood jet and an adjustable plant guard to protect small trees from the chemical. A larger Mk II Arbogard with two flood jets for commercial and forestry work was added in 1972.

The Mk V Allman Rapid barrow sprayer replaced the previous model in the early 1970s. It had a 12 gallon plastic tank, a twin-diaphragm pump driven by a 3 hp Briggs & Stratton engine or an electric motor, a 12 ft spray bar with six nozzles and a hand lance.

Allman introduced a new four-wheel trolley-mounted orchard sprayer and an improved version

4.27. The mid-1970s Allman Midget spraying unit was supplied with a four-stroke engine of an electric motor.

4.28. A hand lance was included in the price of the 15 gallon Allman CR70 barrow sprayer, while a 9 ft wide spray bar was an optional extra.

4.26. The Allman Arbogard was used to ring weed around young trees

of the Midget portable sprayer unit in 1973. The orchard sprayer with a 3 hp Briggs & Stratton engine or optional electric motor had a 55 gallon plastic tank and two hand lances. The Midget had a 75 cc Villiers engine when it was first made in the mid-1960s, but the new model was a self-contained unit with a four-stroke engine, pump, relief valve, pressure regulator and gauge with two sets of hoses and hand lances.

Knapsack sprayers with plastic tanks were still made at Birdbrook in the mid-1980s but after forty years in production run the Rapid sprayer was discontinued in 1987, to be replaced by the new Allman CR70 and

CR100 barrow sprayers. The 15 gallon single-wheeled CR70 and the two-wheeled 22 gallon CR100 had diaphragm pumps driven by a 3 hp Briggs & Stratton engine. A hand lance with 25 m of hose on a reel was standard on both models and a 9 ft spray bar was an optional extra.

Henry Cooper and Frederick Pegler founded Cooper Pegler, a general trading company, in London, WC2 in 1894. Simple insecticide sprayers were eventually included in its stock and successful trading resulted in Mr Pegler obtaining world marketing rights for the French-made Vermorel sprayers. First World War I bombing forced Cooper

Pegler to move to premises in Surrey and later to Burgess Hill in West Sussex. The business was acquired by a Norwegian chemical and plastics manufacturer in 1972 and production continued at Burgess Hill until 1992 when the company moved to Ashington in Northumberland.

In common with other spraying equipment manufacturers, Cooper Pegler made a wide range of pneumatic and pump-operated knapsack and barrow sprayers in the 1940s and 1950s. Its catalogue for 1956 included the No 5 pneumatic hand sprayer which cost £7 9s 6d, the Eclair knapsack sprayer, Presto barrow sprayer and the Ondiver high-pressure sprayer.

The Eclair was used to spray potatoes, ground crops and fruit bushes and to lime wash cowsheds and piggeries. Depending on its intended use the 3½ gallon tank was made of copper, brass or lead-coated copper. Each had a hand-operated diaphragm pump and a compression cylinder inside the tank maintained a steady working pressure to ensure an even output.

4.29. The Cooper Pegler Eclair knapsack sprayer with two low-volume wide-angle nozzles.

The Eclair could be used with a short- or long-reach hand lance and with an angled 3 ft 4 in long lance with two, three or four jets for spraying weeds in a lawn.

The single-wheel Presto barrow sprayer had an 11 gallon lead-coated iron tank, a rubber diaphragm pump suitable for spraying gritty materials and a large compression cylinder maintained a continuous spraying pressure of 60 psi. The Presto was described as an ideal machine for applying fruit tree washes such as tar oil and DDT, lime wash, distemper and creosote.

The Ondiver was a high-pressure sprayer with the choice of an 11 or 22 gallon brass-alloy tank mounted on a two-wheel trolley. It could be used either as a continuously pumped sprayer or as a compression sprayer by filling the compression cylinder with 1¾ gallons of liquid and building up pressure to 45 or 70 psi to spray a group of trees or bushes.

4.30. The Cooper Pegler CP201 pneumatic sprayer with a brass container and pump could be used as a knapsack sprayer or pushed along on a trolley.

The 3¾ hp Cooper Pegler Hurricane Minor and 4½ hp Hurricane Major motorised knapsack sprayers, both with JLO engines were introduced in the early 1980s. With a series of optional accessories the Hurricane could be used to apply liquids, chemical dusts and granules in standard or ultra-low volumes. The Hurricane could also be converted into a flame gun for burning off weeds.

Cooper Pegler products in the late 1980s included the Falcon pneumatic sprayer with a plastic or galvanised steel tank and the CP knapsack sprayer with a plastic tank and a lever-operated diaphragm pump. Cooper Pegler two-wheel barrow sprayers had been replaced by three-wheel trolley sprayers with a 1½ hp two-stroke engine or an electric motor to drive the diaphragm pump. Other features of the trolley sprayers included a 20 gallon fibreglass tank, a hand lance and an 8 ft two-section spray bar. The sprayer could be towed with an optional drawbar after removing the single front wheel.

4.31. The Cooper Pegler Ondiver sprayer was either used as a one-man high-pressure sprayer or as a mechanical sprayer hand-pumped by a second operator.

Most garden and horticultural sprayers were made of expensive but corrosion-resistant copper and brass until 1960 when ICI invited Cooper Pegler to manufacture the world's first knapsack sprayer with a plastic tank. Apart from the hand lance, pump handle, strap buckles and a few bolts the Cooper Pegler Policlair CP3 was made entirely from plastic materials. Polypropylene was used for the pump body and the carrying harness, the four-gallon tank was made from polyethylene and the crankshaft bearings, pump components and hoses were nylon.

The CP3 was the forerunner of a long line of Cooper Pegler knapsack models culminating in the introduction of the Series 2000 CP3 and CP15 sprayers with blow-moulded polypropylene tanks in 1994, 100 years after the two original partners started their business.

4.32. Operator comfort was considered when Cooper Pegler designed the Policlair knapsack sprayer. The tank was shaped to be a comfortable fit on the user's back and sales literature explained that the hand pump was outstandingly easy to use.

4.33. The Dorman Simplex wheelbarrow sprayer, which cost £76 10s 6d in 1951, was used with a spray bar or a hand lance.

Dorman Simplex Sprayer Company products in the late 1940s included barrow sprayers, sprayer kits for Land Rovers and special equipment for spraying potato crops with acid to kill the haulm before the crop was harvested. The company changed its name to the Dorman Sprayer Co in 1951 but production continued at its factory in Ely, Cambridgeshire.

The pump on the 1950 Dorman 15 gallon wheelbarrow sprayer was driven by a 1 hp JAP four stroke engine. The sump was modified to increase its oil capacity which, according to a sales leaflet, gave prolonged periods of trouble-free running. The spray bar covered 7 ft 6 in and an anti drip device on the pump was said to eliminate nozzle drip when the sprayer was shut off. It could be used for a low- or high-volume application to turf and ground crops. The spray bar was secured to the chassis by quick-release clamps, making it easy to remove when changing to the hand lance.

A trailed model with a 15 ft spray bar and a stronger chassis for use with a light tractor or Land Rover was added in 1952. When weed control was required in inaccessible places, the spray bar could be removed, connected to the pump with a suitable length of hose and carried by two people over the problem area.

Knapsack and wheelbarrow sprayers were listed in the Dorman sprayer catalogue for 1961. The Dorman wheelbarrow sprayer had a 15 gallon steel tank and a

bronze and stainless steel gear pump driven by a four-stroke JAP engine. It cost £75 complete with a 7ft 6 in spray bar and five anti-drip nozzles. An optional 10 ft 6 in spray bar was an extra £5 10s 0d and a hand lance with 15 ft of hose cost £4 17s 6d.

The mid-1960s Dorman range included the Osprey Chick knapsack sprayer with a plastic tank for domestic gardens. This was priced at £3 3s 6d while the larger 2 gallon Osprey Continental was £6. There were three models of Dorman wheelbarrow sprayer, comprising the 10 gallon Ely and Wheelaway with engine driven pumps and the Junior Pneumatic sprayer with a 2 gallon tank.

The two-wheeled Ely with a 7 ft 6 in spray bar was mainly used on sports fields while the Wheelaway with a single bicycle wheel and six nozzles spaced across a two-section spray bar was designed for market garden work. The improved Wheelaway II barrow sprayer had a 5 gallon plastic tank carried on both sides of an auto-cycle type wheel with a 75 cc four -stroke engine to drive the gear pump. This delivered chemical to a two-section 9 ft plastic spray bar at a rate of 3½ gallons per minute.

In the mid-1970s Dorman barrow sprayers included the Auto-Spraya and the Osprey Wheelaway, a modified version of the earlier Wheelaway II. The single-wheel Auto-Spraya with a 4 gallon plastic tank and a 6 ft spray bar was designed for spraying fine turf. A lever-type pump driven by the wheel applied chemical at a rate of 10 to 15 gallons per acre.

A barrow sprayer for orchard work with a 22 gallon tank and hand lances costing £520 and the 10 gallon Ely Groundsman with a Briggs & Stratton engine costing £450 were included in the 1978 Dorman catalogue. Ransomes acquired the Dorman Sprayer Co in 1980.

Drake & Fletcher were making Atom and Acorn barrow sprayers in the late 1940s. The Acorn, used for spraying fruit, lime washing, creosoting and disinfecting, was a self-contained unit with a four-stroke engine and a gear pump. Suitable for small orchards, glass houses and private gardens it had an output of 1½ gallons per minute at a working pressure of 75 psi.

Small hand-held and knapsack pneumatic sprayers popular in the 1950s, including those made by Drake & Fletcher, were used by pressurising diluted chemical in a sealed container with a hand pump

4.34. This Dorman wheelbarrow sprayer with a 15 gallon tank and 15 ft spray bar could be hand-propelled or fitted with a tow bar and hitched to a Land Rover or small tractor.

and releasing the contents through a nozzle at the end of a hand lance. The Drake & Fletcher Atom was a pneumatic barrow sprayer with a 1.3 hp JAP engine, air compressor and pressurised tank with a 120 psi working pressure. It was used as an insecticide sprayer for fruit and glass houses. The Atom could

4.35. The Dorman Junior pneumatic sprayer, which cost £32 10s 0d in 1966, was supplied with a foot pump to pressurise the tank contents. A tankful of diluted chemical was sufficient to treat a quarter of an acre of fine turf.

4.36. A pressurised knapsack supplied chemical to the nozzles on this 1960s Dorman band sprayer.

4.37. The Evers & Wall Multicrop sprayer for the David Brown 2D had drop legs with the nozzles mounted in adjustable swivel sockets.

also be used as a compressor for paint spraying, tyre inflation and distempering.

Evers & Wall of Lambourn in Berkshire, another long-established farm and horticultural sprayer manufacturer, made a full range of spraying equipment in the 1950s and 1960s. The Evers & Wall FR2/RCS linkage-mounted rowcrop sprayer for the David Brown 2D, Ferguson 20 and similar tractors was, according to sales information, a versatile machine which enabled one operator to do the work of six men. It was also claimed to commend itself to anyone who grew rowcrops, market gardeners, nurserymen, soft fruit growers and seedsmen.

The FR2/RCS had a 100 gallon tank and the pump could deliver 1,200 gallons an hour at pressures of up to 225 psi to various attachments suitable for all crops from seedling stage to 9 ft high trees. The horizontal spray bar was used with standard non drip fan jets or swivel-mounted twin nozzles on drop legs. The swivel nozzles

could be individually adjusted to give a spray pattern of the required shape, angle and volume. Fruit crops were sprayed with nine swirl jets on two vertical booms carried on an adjustable parallel linkage to suit row widths from 3ft 6in to 9 ft. The spray bars remained upright even when set at their maximum height of 9 ft.

4.38. Market gardeners who owned a Land Rover in the 1950s could use an Evers & Wall sprayer kit to spray their crops.

4.39. The pump for the Evers & Wall Land Rover sprayer was mounted on the front seat and driven by the vehicle's central power take-off shaft.

apply up to 70 gallons an acre from a 7 ft 6 in spray bar at a maximum pressure of 125 psi when travelling at 2 mph. The rate was increased to a maximum of 200 gallons an acre when spraying rowcrops with twin jets on drop legs. Evers & Wall sprayers in the mid-1970s included the Hardi three-wheel barrow sprayer with a 44 gallon plastic tank and a twin diaphragm pump driven by a Briggs & Stratton engine.

The Four Oaks Spraying Machine Co in Sutton Coldfield, Warwickshire, which had specialised in spraying equipment since 1895, included the Excelsior, Kent, Stafford and Ely among the many sprayers listed in their late 1940s catalogues. The Excelsior with a hand pump for spraying, lime washing and disinfecting was used with a bucket or similar container. A hand-held plunger drew the liquid into the pump body and delivered it through a length of hose to a single nozzle on a hand lance.

Evers and Wall also made sprayer kits for the Land Rover and Austin Gipsy in the late 1950s. The tank, supplied either with the kit or a suitable 40 gallon drum, was carried in the back of the vehicle and the spray bar was attached to the front bumper. The pump unit was fitted in place of the middle passenger seat immediately over the vehicle's central power take-off shaft. The unit applied up to 70 gallons per acre at 4 mph and, when it was not needed for spraying, the tank and pump could be used for washing vehicles and buildings.

The Kent pneumatic knapsack sprayer with a 1, 2, 3 or 4 gallon tank was made in the early 1950s. A wooden-handled plunger pump used to pressurise the brass container had a short length of hose connected to a brass hand lance. The four-gallon model weighing 18 lb cost £17 11s 0d in 1959. The Ely narrow-gauge barrow sprayer had an 18 gallon tank with a hand-operated pump, 15 ft of hose and a hand lance mounted on a single wheeled trolley. A sales leaflet suggested that the 12 in wide Ely was ideal for spraying closely spaced rowcrops.

Evers & Wall also made barrow sprayers, knapsack sprayers and spray units for garden cultivators in the 1960s. Knapsack sprayers included pneumatic and hand-pumped models with 2 or 3 gallon brass tanks and hand lances working at 90 to 100 psi. The E & W 10 gallon barrow sprayer on rubber tyred wheels had a powerful brass pump, 10 ft of hose and a hand lance with different jets for lime washing and crop spraying.

Spraying attachments for the Howard-Clifford 700 and Yeoman Rotavators and the Mk I and Mk IV Clifford garden tractors were made by Evers & Wall. They could

4.40. The Four Oaks catalogue explained that its Undentable Syringes would be treasured possessions for those who appreciated a superb syringe.

4.41. *Lime sulphur, creosote and paraffin were among the liquids applied with the Four Oaks knapsack sprayer which cost £10 19s 6d in 1959.*

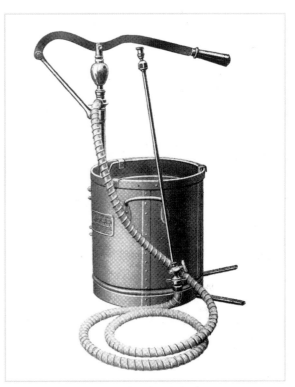

4.42. *The six-gallon Four Oaks Triumph bucket sprayer was made in the 1960s.*

The 1960 Four Oaks catalogue listed about forty different knapsack, bucket, barrow and syringe sprayers, many of which had galvanised steel tanks, brass pumps, hand lances and taps. The Triumph was a typical Four Oaks bucket sprayer with a 6 gallon container, brass pump, 10 ft of hose pipe and a hand lance. The Triumph, the 4 gallon Farmer bucket sprayer at £8 5s 6d and the 10 gallon Rochester with an automatic agitator for £5 1s 0d were used to apply lime wash, creosote, distemper, insecticides, disinfectant and other chemicals.

The Four Oaks Colonial stirrup pump, which cost £5 7s 3d, had a ¾ in diameter brass barrel, a 40 in length of hose pipe, an adjustable nozzle and a polished wooden handle, was an even more basic spraying system. Diluted chemicals or lime wash were applied from the stirrup pump nozzle. The Colonial pump could also be used to extinguish small fires.

The cheaper models of Four Oaks hand-operated and engine-driven barrow sprayers in the early 1960s were supplied with a hand lance, while other more

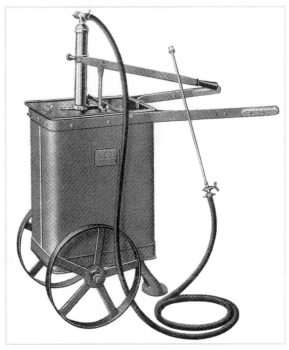

4.43. *The Four Oaks Stafford barrow sprayer had an 18 gallon tank.*

4.44. *A hand pump was used to pressurise the contents of the Four Oaks Streetly pneumatic sprayer.*

expensive sprayers had a hand lance and spray bar. The 6 gallon Victor barrow sprayer for rowcrops was similar to the smaller single-wheel Ely sprayer. It had a double-acting hand-lever operated brass pump and a 2 ft polished brass hand lance with a brass stopcock. With a 25 gallon tank mounted on a two wheeled trolley with handlebars to push it along the Bridgewater was the largest Four Oaks barrow sprayer at the time. Costing £62 10s 0d on pneumatic tyres, it was described as a most efficient and solidly constructed machine, designed for really hard wear under the most trying conditions.

The Four Oaks Senior sprayer, designed for market gardeners and large estates, had a Villiers 70 cc two stroke engine to drive a gear pump which supplied the three-part spray bar or hand lances with dilute chemical from a 15 gallon galvanised tank. The hand-propelled Senior cost £98 10s 0d but growers with spare cash who were not keen to push the sprayer through their crops could buy a 25 gallon self propelled and pedestrian-controlled Four Oaks Spraymobile for £148 10s 0d.

Introduced in 1960, the Spraymobile had a Villiers Mk 12 four stroke engine to drive the pump and propel the sprayer through a clutch and three-speed gearbox. The fully equipped self-propelled sprayer with two inter connected tanks holding 25 gallons of chemical, a hand lance and a 10 ft spray bar with seven nozzles had everything except a seat for the operator.

Spraying attachments were made in the 1950s for the more popular makes of garden tractor and even the

4.45. *The hand-propelled Four Oaks Senior sprayer had an engine-driven pump and the operator was required to push the machine at a constant speed to achieve an even application of chemical.*

Allen Motor Scythe could be used to spray fruit trees and bushes. The Allen Noblox spraying attachment had a piston pump driven by the knife-drive crank, suction and delivery hoses, pressure regulator and a hand lance. The pump was bolted to the cutter bar bracket and with the tank on a load-carrying platform above the engine or in a trailer, the Allen scythe became a self propelled spraying unit.

4.46. The Four Oaks Spraymobile was a self-propelled version of the Senior sprayer.

A sprayer attachment was included in the 1950 Clifford Aero & Auto cultivator catalogue. The pump was attached to the rotary cultivator drive shaft and the tank was close-coupled under the handlebars. Four nozzles were arranged around the tractor and a connection was provided for a hand lance. Each nozzle had a control tap and when using a single nozzle the pump, working at a pressure of between 350 and 400 psi, emptied the 18 gallon tank in twenty-one minutes. With four jets in use the pressure was reduced to 150 psi and the tank was emptied in six minutes having travelled little more than 200 yards. A later model with a boom and hand lance introduced in 1954 for the Clifford Mk IV was attached to a cultivator with two bolts.

A similar outfit was made for the Allen & Simmonds Auto Culto Model M garden tractor in the early 1950s. The pump emptied the 40 gallon tank in sixteen minutes and the 350 psi working pressure was sufficient to operate two hand lances.

The Coleby mobile sprayer was originally made as an attachment

4.47. Spraying with an Allen Motor Scythe.

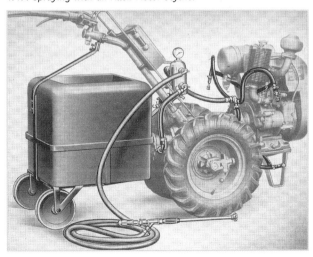

4.48. The Clifford sprayer attachment.

4.49. The Coleby Junior sprayer.

Ransomes MG crawler owners had a wide choice of sprayers with power take-off driven pumps. Ransomes' Mk I Cropguard Junior with a 30 gallon tank mounted on the MG's hydraulic linkage had a roller vane pump, a 17 ft spray bar and optional hand lances for orchard spraying.

A & G Cooper of Wisbech, Cambridgeshire; W Weeks & Son at the Perseverance Works in Maidstone, Kent, and Drake & Fletcher also of Maidstone made trailed sprayers for the MG. The pump on the 100 gallon Cooper Demon Z200 delivered 3½ gallons per minute to two hand-held lances at a maximum pressure of 400 psi. The pump was mounted on a bracket bolted to the tractor and driven by a roller chain from the power take-off.

for Coleby garden tractors. It had a 30 gallon tank and a three-cylinder pump driven by a 3½ hp JAP 4/3 four stroke engine with an output of 120 gallons per hour. The maximum working pressure was 400 psi with a single nozzle and 250 psi when using a twin-nozzle hand lance.

The trailed Drake & Fletcher LO Estate sprayer had a 60 gallon tank and the pump, also mounted on the tractor and belt-driven from the power take-off, delivered up to 6 gallons per minute at a maximum pressure of 250 psi. A hydraulic agitator kept the contents of the tank well mixed. The LO Estate model could be used with a 12 ft spray bar, hand lances, adjustable nozzle guns for orchard spraying or a special spray bar with fifteen nozzles for hop gardens.

The sprayer pump on the trailed Weeks Model M driven by a universal drive shaft from the power take-off delivered 4 gallons per minute at 200 psi. The pump and 50 gallon tank were mounted on a two-wheel tubular chassis on pneumatic-tyred wheels. The Model M was used to spray ground crops or soft fruit and hand lances were provided for orchard work.

Small mounted sprayers made in the 1940s 1950s for the Ferguson TE 20, Allis Chalmers Model B and similar tractors were suitable for smallholdings and market gardens. Most farm tractors and sprayers were too big for this type of work by the late 1960s and this led to the

4.50. The Coleby mobile sprayer unit.

introduction of a new generation of small mounted sprayers for compact tractors.

Soon after Lely Import became the UK distributor for Iseki tractors it introduced the Conquest 88 mounted sprayer for Iseki, Kubota and other makes of compact tractor. The Conquest sprayer with an 88 gallon plastic tank had many of the features found on

farm tractors at the time, including a twin diaphragm pump, a stainless steel spray bar and colour-coded plastic nozzles.

Developments in the domestic sprayer market included the Walkover sprayer, invented by Paul Ridgeon for the application of herbicides and liquid fertilisers to lawns. Launched in 1980, the Walkover had a land-wheel driven pump to deliver chemical from a small plastic tank to a series of nozzles placed close to the ground at the front of the machine. The pump only supplied chemical when the sprayer was pushed in a forward direction and the supply ceased when it was stationary or pulled backwards.

Allen Power Equipment acquired Walkover sprayers in 1990 and added larger models for gardens, nurseries and sports grounds. The

4.51. The power take-off driven pump on the Weeks Model M sprayer for the Ransomes MG crawler had an output of 250 gallons per hour at 200 psi pressure.

4.53. The late 1980s Walkover Groundsman sprayer was used to control weeds in fine turf.

4.52. Orchard spraying with the Cooper Demon sprayer was done with 6 ft hand lances and 3 ft lances were recommended for bush fruit and in glasshouses.

4.54. The Sprolley dual-purpose sprayer was made in the mid-1990s.

Fieldmaster trailed sprayer with a 12 gallon tank and a 6 ft spray bar was made for small ride-on tractors and had a work rate of about 14 acres a day.

L E Toshi Ltd, the first sellers of the Iseki compact tractors in the UK, introduced Sprolley, a dual-purpose sprayer on a lightweight trolley with a spray bar and hand lance in 1994. Later marketed by Sisis, the Sprolley had a diaphragm pump driven by an electric motor supplied with current from a 12 volt battery which also powered the front axle on an alternative self-propelled version of the machine.

CROP DUSTERS AND MIST BLOWERS

Insecticide and fungicide powders and dusts have been used to treat horticultural crops at very low application rates for many years. Various types of

powder applicator or duster, which issued a penetrating cloud of fine dust, were made in the 1940s and 1950s. They ranged from small hand-held dust guns for gardens and greenhouses to dusting attachments for garden tractors. Most were made with little consideration for the health and welfare of the user.

Several companies, including Allman, Cooper Pegler and Drake & Fletcher made hand-operated dusters for gardeners and smallholders. Drake & Fletcher introduced the 'Dustejecta' and the New Armada dust gun in 1949. The New Armada dust gun and mistifier was carried on the operator's chest and the dust was blown from either a single or double outlet by a stream of air from a hand-cranked fan.

A mid-1950s Cooper Pegler catalogue included

hand bellows dusters and a knapsack powder duster and the Rotver dust gun. Cooper Pegler's hand bellows dusters with a 2 lb capacity container had a hand-operated agitator to supply powder to an air stream created by pumping the bellows with a lever. The knapsack duster with a 15 lb container carried on the operator's back had a semi-rotary handle to work the double-acting bellows which distributed sulphur and other powders through a hand lance. A double outlet lance was used when dusting two rows of potatoes at a time.

The Cooper Pegler Carpi duster, with hand-operated bellows, cost £2 17s 6d in the late 1960s. The user was required to pump the hand bellows, which blew the dust from a spreader nozzle at the end of a 3 ft tube, while walking along a row of plants.

4.55. The Cooper Pegler Rotver dust gun with a lead-coated sheet-iron container, a single or double outlet lance hand and a hand-cranked fan ran at 2,500–3,000 rpm.

The Allman HandOp hand-cranked dusting machine, introduced in 1946, was carried on the user's chest with fully adjustable shoulder straps to hold it in place. The HandOp weighed 21 lb with a full hopper of powder, which was distributed from a nozzle by an air blast from a hand-cranked six-blade fan. Sales literature stated that the HandOp would dust ground crops, bushes or trees with equal efficiency and that it was light, strong, perfectly balanced and easy to operate for long periods without undue fatigue. The technical details explained that the 55 rpm fan had an output of 47½ cu ft of air per minute at a velocity of 2,500 ft per minute.

The Allman Speedesi, the first power duster made by Allman, was awarded a Silver Medal for new implements at the Royal Agricultural Society's Show at Lincoln in 1947. The Speedesi, which could be used with various farm and garden tractors including the Ransomes MG 2, used the engine exhaust gases to distribute powdered pesticides on to the crop. As well as carrying the powder to the

4.56. The Allman HandOp dusting machine with a hand-cranked six-blade fan introduced in 1946 was strapped to the operator's chest.

4.57. Sales literature described the Allman Pestmaster Minor, made between 1946 and 1960, as an all-in-one machine suitable for insecticides, pesticides and powdered fertilisers.

spreader outlets, the exhaust gas also pre-heated the dust before distribution, and sales literature took full advantage of the well-known fact that warmed powder was more effective in the extermination of pests. The Speedesi could be used to apply dust to ground crops including potatoes and celery from pendant nozzles spaced across a 10 ft wide boom and in orchards with two fixed fish-tail nozzles or a long-handled dusting lance.

The Allman Pestmaster Minor duster had a single bicycle-type wheel, tubular handlebars, a powder hopper, a feed mechanism and a centrifugal fan with delivery pipes. Two small brushes in the hopper agitated the powder and swept it at a controlled rate through an adjustable-sized outlet into a flow of air created by a wheel-driven fan. The powder was blown through the delivery pipes to fish-tail spreaders or curved deflector plates.

The Allman Midget duster, with a 34 cc two-stroke engine and fan, was attached with straps to the user's chest and the powder hopper was carried on the back. The fuel tank contained enough petrol and oil mixture for 1½ hour's work and the hopper held 20 lb of dust. An alternative model required the engine, fan and powder hopper to be strapped to the user's chest. It was suggested that the more enthusiastic person could treat two rows at the same time by attaching a two-way branch and twin flexible dusting pipes to the fan housing. Allman also made one and two wheeled barrow versions of the Midget duster.

The Drake & Fletcher Motorised Dust Gun had a crankshaft-mounted fan sandwiched between a 6 lb capacity hopper and a two-stroke engine. The dust

4.58. The Drake & Fletcher Armada dust gun.

gun with a full hopper weighed about 24 lb and the operator, who was required to carry the machine and direct the nozzle at the target, could apply up to 60 lb of powder per acre.

Duster attachments were also made for several models of garden tractor in the 1950s. The recirculation duster for the Howard Bantam had a centrifugal fan driven by the cultivator rotor shaft. The powder was released into an air stream from the fan and discharged from three nozzles under a canvas canopy to create an intense cloud of chemical dust. Some of the dust adhered to the plants and the rest was sucked back into the fan housing where it mixed with fresh powder from the hopper. The fan was vee-belt driven and the feed agitator in the hopper was driven by a worm-and-wheel gear arrangement. Application rates of 40 to 120 lb per acre were possible and the amount was regulated by selecting any one of four worm gears to drive the powder agitator.

Potentially health-threatening hand-operated dusters were replaced with combined dusting and misting equipment in the late 1950s. The new equipment was engine driven and carried knapsack fashion on the operator's back.

The Drake & Fletcher Mamba could be converted from a duster to a mist sprayer in a minute or so by swopping the dust hopper for a tank. A 75 cc engine with crankshaft-mounted fan was used to apply mist at rates of a few pints per acre to control pest infestations or up to 20 gallons per acre in orchards. The spray mist or powder was discharged from a hand-held nozzle on a flexible hose and directed at the target by the user.

Kent Engineering & Foundry made the KEF Motoblo combined knapsack mist blower and duster in the mid-1970s. It weighed 53 lb with its 3½ hp engine and a full 2½ gallon polythene chemical tank and could apply up to 6 pints per minute from a hand-held discharge nozzle. The smaller Motoblo Junior with a 2 hp engine also

4.59. The nozzles under the canopy of the Howard Bantam recirculation duster created a dense cloud of chemical dust.

applied up to 6 pints of chemical per minute and with a full tank it weighed 36 lb.

The 1970 Allman catalogue included a knapsack L80 Mistblower/Duster which, with a full set of accessories, could be used for misting, dusting and flame weeding. The basic price at the time was £65 but by 1977 inflation had increased the price to £185 and the flame thrower for killing weeds was still included in the list of optional extras.

4.60. The Allman duster for the Farmers' Boy light tractor had a brush-feed mechanism to meter powder from the 24 lb capacity hopper into a stream of air.

A 1948 Ransomes sales leaflet explained how applying insecticide dusts with the Ransomes flea beetle duster would control flea beetle and turnip fly on brassica seedlings. The duster was pushed by hand and a land-wheel driven roller feed mechanism applied the dust in a 6 in wide band over each row of young plants at a rate of 35 to 45 lb per acre.

Equipment for applying chemicals at rates as low as 1 pint per acre became available in the mid-1950s. The principle of ultra-low volume spraying with a spinning disc to atomise liquids into minute droplets was developed by Edward Bals, who founded Micron Sprayers in 1953. It was found that tiny quantities of liquid, evenly atomised and distributed, controlled pests equally as well as the more traditional application rates with conventional sprayers.

In 1970 very low-volume application progressed a stage further with the introduction of a hand-held Ultra Low Volume Applicator. The Ulva sprayer applied 70 micron droplets from a small spinning disc driven at 6,000 rpm by a 1½ volt electric motor. When spraying outside, the wind was utilised to carry the minute droplets of chemical to the plants but the more expensive Ulvafan was required for ultra-low volume applications to glasshouse crops. This hand-held unit had an electric motor-driven fan behind the atomiser to create the necessary air flow to carry the chemical to its target.

Other types of ultra low-volume applicator made during the 1980s included a hand-held twin atomiser with a knapsack chemical container and the single-wheeled Micronette barrow applicator with a 1 gallon tank and a small petrol engine to drive the fan and atomiser disc.

4.61. The Cooper Pegler Carpi dust gun had hand bellows to create the air stream used to distribute the powder.

4.62. The Ransomes flea beetle duster was used for the control of Flea Beetle, Black Bob or Turnip Fly.

Chapter 5

Lawn Mowers

Early Days

Apart from the gardens around stately homes and in public parks, a closely trimmed lawn remained a rare sight until the late 1800s. Teams of three or four men usually cut lawns with scythes, a task requiring considerable skill to achieve a good finish. Daisy heads were chopped off between each mowing with a daisy rake. The first cylinder lawn mower was made by Edwin Budding in 1830 but many years passed before the lawn mower became a common sight throughout the land.

This chapter deals with the more recent history of lawn mowers but it would be incomplete without a brief account of the development of grass cutting machinery. Edwin Budding was an engineer responsible for machines used to shear the nap from cloth in textile factories who conceived the idea of using a similar mechanism to cut grass. Budding went into partnership with John Ferrabee at Stroud in Gloucestershire and the first Budding-designed lawn

mowers were made by Ferrabee at his Phoenix foundry in 1831.

Two years later J R & A Ransome in Ipswich, Thomas Green in Leeds and Alexander Shanks in Arbroath were all granted licences to manufacture the Budding mower. It was gear-driven from the rear roller and had a flat box at the front to collect the grass cut by the 21 in wide cylinder. Ransomes improved the design from time to time and by 1852 the company had made about 1,500 mowers in Ipswich. Ransomes stopped making the Budding mower in 1858 and from that date supplied its dealers with machines bought from Thomas Green and Alexander Shanks.

5.1. The patent specification for Edwin Budding's 21 in cut lawn mower stated that country gentlemen would find using this machine an amusing, useful and healthful exercise.

5.1A. Ransomes introduced the Budding patent cylinder lawn mower in 1833.

5.2. The Lancashire Steam Motor Co introduced its pedestrian-controlled steam-powered mower in 1893.

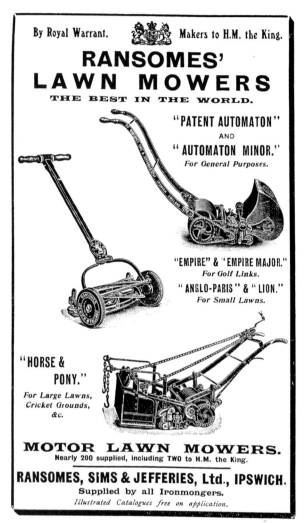

5.3. Ransomes advertised its side-wheel, roller and pony mowers in the Flower Garden magazine in 1903.

Follows & Bate made the first side-wheel mowers in the late 1860s. They were cheap to buy and four models with 6 to 10 in wide cutting cylinders soon became popular with the gardening public.

The Pennsylvania side wheel mower, made in Philadelphia, was sold in Britain by Lloyd Lawrence & Co of London, EC2, which later became Lloyds of Letchworth in Hertfordshire. The Vicar of Booton in Norfolk, one of the many Pennsylvania hand mower owners, wrote to the company in the early 1900s complimenting it on its machine, which had exceeded all his expectations.

Alexander Shanks made a 27 in cut lawn mower in 1841 which was light enough for a small pony to pull without damaging the grass with its hooves. The phrase 'to travel by Shanks's Pony' is said to date back to the days when people walked behind a Shanks pony mower. Shanks also made a 42 in cut horse-drawn mower developed from the earlier pony mower. The Ransomes, Sims & Head catalogue for 1870 included the 'Horse Power' lawn mower with cutting widths from 30 to 48 in for large lawns, cricket grounds and tennis courts. Pony mowers with 26 and 30 in cutting cylinders were added a year or so later. The horse wore leather boots to minimise damage to the turf and in the late 1890s a set of four boots cost £1 5s 0d.

The Lancashire Steam Motor Co made the first steam-powered, pedestrian-controlled lawn mower in 1893. It weighed 1½ tons and took ten minutes to get steam up from cold before it was ready for use. The Lancashire Steam Motor Co eventually became Leyland Motors. Alexander Shanks introduced a steam-driven ride-on roller mower in 1900. Claimed to be more efficient than a pony mower, it had the added advantage of ensuring no damage from horses' hooves.

Ransomes, Sims & Jefferies made the world's first commercially manufactured petrol-engined mower in 1902. Designed by James E Ransome, the 42 in cut cylinder mower with a 6 hp water-cooled engine enjoyed considerable success. The first one was sold to a Mr Prescott-Westcar at Rochester and the

5.4. Ransomes made the world's first commercial motor mower in 1902.

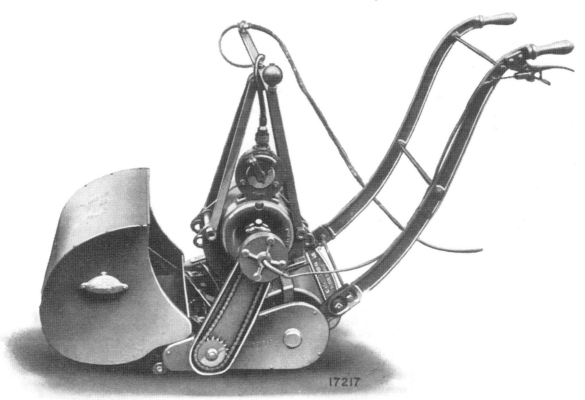

5.5. The mid-1930s 14 in cut Ransomes Electra with a 1½ hp mains electric motor cost £27 10s 0d.

I make short work of long grass

However long the grass the HAYTER 24in. MOTOR SCYTHE will cut it quickly and neatly, with little effort, and leave no long stalks. It will cut grass not possible with a mower. Lightweight, highly manœuvrable, very strongly constructed and cheap to operate.

Hayter
MOTOR SCYTHE

£48 . 15 . 0

POWERED BY
Villiers
ENGINES

HAYTERS (SALES) LTD
Spellbrook Lane, Bishop's Stortford

5.6. An early advertisement for the Hayter Motor Scythe.

second was sold to Cadburys for its Bournville sports grounds. Within a year Ransomes were making 32 and 42 in ride-on mowers and 2½ hp pedestrian-controlled machines with 24 and 30 in cutting cylinders.

Thomas Green of Leeds introduced a 24 in cut motor mower in 1903 but in spite of strong competition from other manufacturers Ransomes sold more than 600 motor mowers over the next ten years.

Charles H Pugh introduced the first motor mower for the home garden in 1921 and for the next twenty years several manufacturers developed small pedestrian-controlled roller and side-wheel motor mowers for the private garden. They included Arundel, Atco, Coulthard & Co (Presto), Dennis Brothers, Greens, J P Engineering, Royal Enfield, Qualcast, Ransomes, Shanks, Suffolk Iron Foundry and H C Webb.

The design for a horse-drawn gang mower was patented by a Mr Worthington in America in 1914 and Ransomes made its first Worthington gang mowers under licence in 1921. A second Worthington patent was taken out in America for a motorised triple golf green gang mower and Ransomes used this design for the Overgreen mower it first made under licence in 1937. The Overgreen equipped with three Certes fine turf hand mowers made it possible for one man to cut eighteen golf greens in a day.

Ransomes claimed another first in 1926 when it introduced the Electra mains-electric cylinder mower for private gardens, and the Bowlic mains-electric mower for bowling greens appeared a year or so later. The Rotoscythe, the world's first rotary mower, was made by Power Specialities at Maidenhead in 1933. Douglas Hayter entered the rotary mower market in 1947 and within a few years rotary mowers were being sold in far greater numbers than cylinder mowers. The first commercial Webb battery-powered electric mower appeared in 1959 and the Flymo rotary hover mower arrived on the gardening scene in 1964.

The scissor-cutting action of reciprocating knife mowers had been used in agriculture for almost a century before it appeared on the horticultural scene in the 1930s. The best-known self propelled Allen scythe was made for nearly forty years, while similar pedestrian-controlled machines with an engine-driven cutter bar were made by several companies including Atco, Lloyd, Mayfield and Teagle. Cutter bar mowers were also made for some models of two wheel garden tractor. Hundreds of different lawn mowers have appeared since the 1930s and the following pages provide little more than a snapshot of the manufacturers and the machines they have made during the last sixty years.

Hand Mowers

ATCO

Charles H Pugh, who pioneered the motor mower for the private gardener, owned the Atlas Chain Co. The Atco brand name was derived from the first and last two letters of the company name. Qualcast acquired Charles H. Pugh in 1962 and although Atco was better known for petrol-engined and electric motor mowers it also sold Atco hand mowers in the early 1960s.

The 12 in cut Atco hand mower with a six-blade cylinder and tubular handlebars cost £13 5s 0d complete with a grass box in 1969. Atco hand

mower production was discontinued in 1972. Qualcast moved to Stowmarket, Suffolk in 1975 and introduced the Atco Elite side-wheel mower in 1980. The 14 and 16 in cut Elite with a five-blade cylinder had large diameter aluminium wheels with solid rubber tyres and a large capacity rear grass box.

Many gardeners bought a small electric hand mower in the early 1980s and with declining sales the last Atco Elite hand mowers were made in 1984. Atco returned to the hand mower market again in 1989 when the new 14 and 16 in cut Super Clipper side-wheel mowers were announced. Limited numbers of the Atco Super Clipper with a five-blade cylinder, heavy-duty plastic grass box and tubular handlebars were made for the next six years.

FOLBATE

Follows & Bate of Gorton near Manchester, which made hand-propelled and pony cylinder mowers in the mid-1850s, patented the principle of the side-wheel lawn mower in 1869. This was an important development because, unlike the cylinder mower with a roller which pressed down the grass before it was mown, the Folbate Climax side-wheel mower was able to cut standing grass. Internal teeth on the rims of the side wheels drove the cutting cylinder and a built-in ratchet mechanism allowed the wheels to turn at different speeds. Made in four cutting widths from 6 to10 in the Climax had a small wooden roller behind the cylinder. A metal grass box could be attached at the front or rear of the mower.

Follows & Bate were making the Falcon ball bearing roller lawn mower when Qualcast bought the company in 1938. The Falcon, which cost £2 8s 6d, had a six-blade 12 in wide cutting cylinder, free-wheel drive and a hand wheel to regulate the cutting height. Qualcast, which used the Folbate brand name until 1966, made very few changes to the existing range of Folbate mowers. The Qualcast catalogue for 1955 included 10 and 12 in cut J1 Folbate side-wheel mowers with cast-iron wheels and

5.7. Atco made Super Clipper side-wheel mowers from 1989 to 1995.

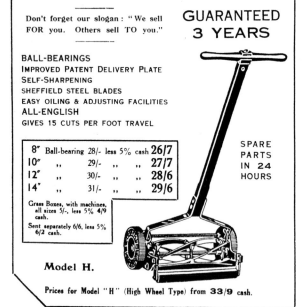

The Ever Popular BALL BEARING MODEL "E"

The "QUALCAST" Model "E" has become so firmly established in public favour that we have decided to adhere to the popular specification, incorporating ball bearings on all sizes; and we again emphasize how important it is for you to order early and in good quantities.

The public know what they want and will not be put off with substitutes, neither will they wait.

If they cannot obtain their requirements at the first enquiry they will go to a dealer who has "QUALCASTS" in stock.

BE SURE to order EARLY and ENOUGH.

Don't forget our slogan: "We sell FOR you. Others sell TO you."

GUARANTEED 3 YEARS

BALL-BEARINGS
IMPROVED PATENT DELIVERY PLATE
SELF-SHARPENING
SHEFFIELD STEEL BLADES
EASY OILING & ADJUSTING FACILITIES
ALL-ENGLISH
GIVES 15 CUTS PER FOOT TRAVEL

8"	Ball-bearing	28/-	less 5%	cash	26/7
10"	,,	29/-	,,	,,	27/7
12"	,,	30/-	,,	,,	28/6
14"	,,	31/-	,,	,,	29/6

Grass Boxes, with machines, all sizes 5/-, less 5% **4/9** cash.
Sent separately 6/6, less 5% **6/2** cash.

SPARE PARTS IN 24 HOURS

Model H.

Prices for Model "H" (High Wheel Type) from **33/9** cash.

5.8. Features of the Follows & Bate S2 side-wheel mower included a handle and roller of the finest-quality varnished wood, 7 in diameter wheels with wide rims and a cutting cylinder with five Sheffield steel spiral blades.

FOLLOWS & BATE'S "FALCON"

BALL BEARING ROLLER LAWN MOWER

More Blades to cut the Blades!

GUARANTEED FOR THREE YEARS

BRITISH AND BUILT BY SPECIALISTS

because it has

Six Cylinder Cutters — Giving a smoother finish to your Lawn.
Free Wheel Drive to the Cutting Cylinder, making it easy to operate.
Handles Adjustable for height.
Bush Roller Chain Drive Completely Enclosed.
Hand Wheel Regulation of Cutting Cylinder and height of cut.
Frame, Roller and Cylinder Bosses of Unbreakable Pressed Steel.
Practically Indestructible.

48/6 NETT CASH

one size
12 inch cut

COMPLETE WITH LARGE CAPACITY GRASS BOX

Obtainable from Ironmongers and Stores. In case of difficulty send price direct giving name of local Ironmonger

FOLLOWS & BATE LTD.,
GORTON,
MANCHESTER 18.

5.9. The Folbate Falcon was guaranteed for three years and according to advertisements it was virtually indestructible.

GREEN

Thomas Green, a Leeds blacksmith, who in 1833 was granted a licence to make the Budding roller mower went on to make the first hand-propelled roller mowers of his own design in 1855. He introduced the patented Greens Silens Messor roller mower in 1859 (Silens Messor meaning silent running). The new model had the distinction of being the first chain-driven lawn mower. Silens Messor hand and pony roller mowers were made until the mid-1930s. The Green Prince roller mower and New Century side-wheel mower were introduced in 1929 and the side-wheel Green Clipper which only cost £1 5s 0d appeared in the early 1930s. The mid-1930s Green's Defiance roller mower with a 10 in wide six-blade cylinder complete with a grass box cost £2 15s 0d.

wooden handle and the 12 in cut F1 side-wheel mower with solid rubber tyres. The 10 in cut J1 cost £3 6s 4d, the 12 in model was £3 9s 4d and an optional grass box was an extra 10s 10d. The 12 in cut Falcon roller mower complete with grass box cost £5 plus 18s 3d purchase tax. About 1,500 Suffolk Falcon roller mowers were sold in 1970 but due to declining sales the last Falcon mowers were made in 1972.

5.10. Green's Clipper side-wheel mower cost £1 5s 0d in 1939.

5.11. This Qualcast side-wheel mower advertisement appeared in 1938.

QUALCAST

The history of Qualcast dates back to 1801 when the Jobson Foundry Co, based in Sheffield, was making stoves and grates. The company moved to Derby in 1849 and shortly after changed its name to the Derwent Foundry Co. The first Qualcast lawn mowers were made at Derby in 1920 and before long the Derwent Foundry Co was selling a range of side-wheel and roller mowers at prices below those of most other British-made machines. The Derwent Foundry Co was a major lawn mower manufacturer by 1928, when it became a public company and the name was changed to Qualcast, a derivative of Quality Castings.

The first hand-propelled Qualcast Panther roller mowers were made in Derby in 1932 and within six years more than a million of them were in use. The

1939 Qualcast price list included four Model E side-wheel mowers with 7 in diameter wheels and 8 to 14 in wide cylinders and four Model H machines with 9 in diameter wheels and cutting widths of 10 to 16 in. Prices ranged from £1 8s 0d to £1 18s 6d. The Qualcast 12 in cut Panther roller mower complete with a grass box was £2 15s 0d.

Qualcast acquired Follows and Bate in 1938 but the onset of war brought lawn mower production to a temporary halt. Munitions including mortar bombs and hand grenades were made at the Victory Road factory in Derby during the war years.

209

Limited production of Qualcast Models E and H side-wheel mowers and the Panther was resumed at Derby in 1946. A 1950 magazine advertisement informed readers that the self sharpening Panther, which ran on ball bearings and had a simple 'click' adjustment for the cutting cylinder, was the world's most popular mower with more than three million satisfied users.

A new 12 in cut Qualcast Model B1 side-wheel mower with a light tubular steel handlebar introduced in 1949 was a somewhat revolutionary change from the traditional Tee shaped varnished wooden handle. Solid rubber cross-tread tyres were another innovation claimed to give a perfect grip with a silent and easy action, which reduced friction to a minimum. A later version of the 12 in cut Model E side-wheel mower was advertised in 1950 at the reduced price of £3 19s 10d plus 17s 5d purchase tax. An improved 14 in cut H1 side-wheel mower with solid rubber-tyred wheels and a tubular steel handle with rubber handgrips appeared in 1951.

With ever-increasing sales of motor mowers in the late 1950s the Qualcast range of hand mowers had been reduced to the tubular steel-handled B1 and E1 side-wheel machines and the 12 in cut Panther. Qualcast dropped the Folbate brand name in 1966 and the earlier Folbate Falcon roller mower was sold as the Suffolk Falcon until 1972.

The Suffolk Iron Foundry became part of the Qualcast Group in 1958, as did Charles H Pugh in 1962. The Qualcast Group then merged with Birmid Industries in 1967 to form Birmid Qualcast. When Atco and the Suffolk Iron Foundry amalgamated in 1969, they remained within the Birmid Qualcast group and

5.12. The Qualcast Panther cost £2 12s 3d in 1938. Twelve years later it was £7 2s 6d.

5.13. The Qualcast Model E was made with 8 to 14 in wide cutting cylinders.

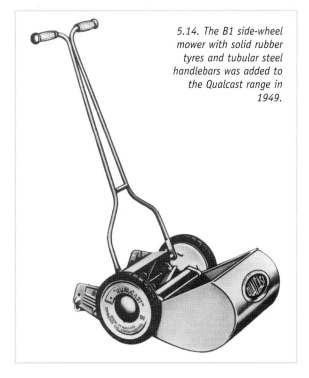

5.14. The B1 side-wheel mower with solid rubber tyres and tubular steel handlebars was added to the Qualcast range in 1949.

5.15. The Qualcast Superlite Panther.

became Suffolk Lawn Mowers. Production of Atco mowers was transferred to Stowmarket, Suffolk in 1975.

Birmid Qualcast widened its interests in 1984 with the acquisition of Wolseley Webb and Landmaster. Birmid Qualcast was then bought by Blue Circle Industries in 1988. The Derby factory was closed in 1991 with mower production concentrated at Stowmarket. Blue Circle then disposed of its garden products division in a management buyout in 1992 and the new owners adopted Atco Qualcast as the company name. Atco Qualcast is now part of the Bosch organisation.

The Qualcast Super Panther superseded the Panther roller mower in the early 1960s while the E1 side-wheel mower was discontinued in 1967 after a production run of almost thirty years. Qualcast mower design moved forward in 1969 with the launch of the 12 in cut Superlite Panther which cost £8 19s 6d and carried a ten-year guarantee. Diecast aluminium side plates and tubular steel handles with plastic grips reduced the weight of the new model which had dirt seals on the bearings, caps over the oil holes, an enclosed chain drive and twin nylon 'click' blade adjusters.

The popularity of the motor mower in the late 1960s hastened the decline of the hand mower. The Qualcast Superlite Panther and 12 in cut Model Q7 side-wheel mower, which replaced the B1 in 1973, were the only hand-propelled mowers listed in the 1974 Suffolk Lawnmowers catalogue. The Qualcast Q30 side-wheel mower cost £29.95 when it superseded the Q7 in 1980. The Q30 cylinder had five self sharpening steel blades with self lubricating bearings and rubber-tyred wheels threw the clippings rearwards into a grass collector.

The new hand-propelled Panther 30 DL roller mower with rollers at the front and rear and a five-blade cutting cylinder appeared in 1980. With many gardeners buying a new mains-electric mower, sales literature for the Panther pointed out that one of its advantages was the lack of a trailing cable which could get tangled up in trees or bushes. The Panther 30 and Panther 35 side-wheel mowers with a rear grass collector were added in 1981. The model numbers referred to the 30 cm (12 in) and 35 cm (14 in) cutting widths. The original Panther was made in 1932 and sixty years later the name survived in the shape of the 12 in cut Panther 30 S and 30 DL roller

5.16. The Qualcast Panther 30 side-wheel mower with a 12 in cutting cylinder was still being made in the mid-1990s.

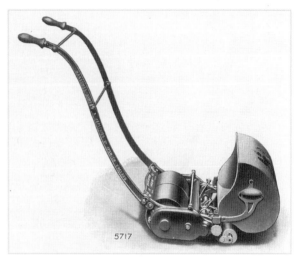

5.17. The first Ransomes Ajax mowers were made in 1933.

5.18. Ransomes Anglia roller mowers were made in the late 1930s.

mowers and the side-wheel Panther 30 with a rear grass collector. Advertisements claimed that the Panther with its pivoting handle and large diameter rear roller left a beautifully striped lawn.

RANSOMES

Ransomes traded under various names in its 200-year history. In 1833 J R and A Ransome made the Budding gear-drive lawn mower; in 1846 Ransomes & May formed a partnership which in 1852 gave way to Ransomes & Sims. Ransomes Automaton roller mowers with 10 to 18 in wide cutting cylinders appeared in 1866 and a new partner joined the business in 1869 to form Ransomes, Sims & Head. Following a brief spell as Ransomes, Head & Jefferies

from 1881, a private limited company trading as Ransomes Sims & Jefferies was established in 1884, becoming a public company in 1911.

In the early 1930s Ransomes made at least ten different roller and side-wheel mowers. The list included the Anglia, Ransomes de luxe Centenary, Ajax and the Certes for fine turf, while side-wheel mowers included the Leo, Cub, Coronet and the Kutruf with a four-blade cylinder for cutting rough grass up to 7 in high. Some side-wheel machines were made with a single six-foot-long handle. Called bank mowers, they were used to cut sloping areas of grass and cost between 2s 0d and 2s 6d more than the standard model.

Ransomes' side-wheel and roller mower catalogue

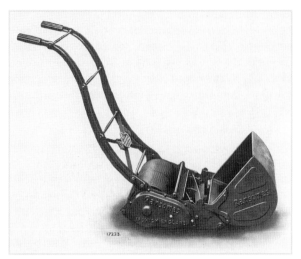

5.19. The RSJ de luxe was introduced in 1932 to commemorate 100 years of Ransomes lawn mowers.

5.20. Ransomes side-wheel mowers with long handles were called bank mowers.

for 1938 included the Ace, Cub and Leo side wheel models. With prices starting at £2 1s 3d they were described as admirable machines for small lawns. The Atlas and Anglia roller drive mowers, which cost from £3 8s 9d complete with grass box, and the Ajax - introduced in 1933 - were said to be best suited to medium-sized areas of grass.

With the exception of gang mowers used to cut airfields, production of Ransomes lawn mowers came to an almost immediate stop in 1939 when the factory turned to the production of armaments and aircraft components. The side-wheel Ripper was one of the first new models made at Ipswich after the war years. Introduced in 1946, the 14 in cut Mk I Ripper was designed to cut grass up to 7 in long. An improved Mk II Ripper replaced the Mk I in 1963 and was made until 1974.

Mk I and Mk II Ajax mowers were made between 1933 and 1939. Production restarted in 1946 and the Ransomes 1950 mower catalogue included the Ajax, Astral, Mk III and Certes roller mowers together with the Ariel, Moon, Tiger and Ripper side-wheel mowers. The Mk IV Ajax appeared in the late 1950s when Ransomes' publicity material suggested that

5.21. The Ransomes Ripper side-wheel mower for rough grass was made from 1946 to 1974.

keen gardeners would be proud to own the new lightweight and easy-to-use Mk V Ajax. It was further suggested that Ajax owners would be even more proud of the velvety two tone finish it would leave on their lawns. The popular Ajax roller mower was made for more than thirty years before being discontinued in 1970.

The Conquest, made from 1959 to 1965, was a sophisticated-looking 12 in cut side-wheel mower with a roller chain driven six-blade cylinder under a streamlined sheet metal cover. The 14 in Ripper side-wheel mower and the long-established Certes with a 16 in, ten-blade cylinder for cutting fine turf completed the 1960s range of Ransomes hand mowers. The Conquest, which cost £7 11s 0d complete with a canvas grass catcher, and the Mk V Ajax priced at £12 5s 0d were the only Ransomes hand mowers in 1965.

With a change of policy in the mid-1970s Ransomes discontinued its range of domestic mowers to concentrate on the production of professional grass care machinery

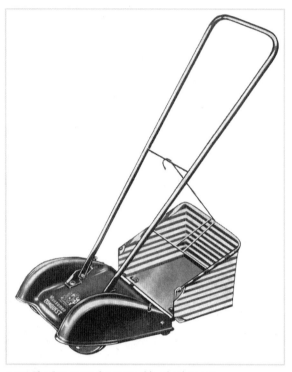

5.22. The Ransomes Conquest side-wheel mower.

5.23. The Mk V Ajax was still being made in 1969.

SHANKS

Shanks, along with Ransomes and Thomas Green, were the major lawn mower manufacturers in the late 1920s. The company advertised mowers for all purposes with prices starting at £2 5s 0d. The Shanks Ivanhoe mower with an eight-blade cylinder for fine turf was introduced in 1927 and the 12, 14 and 16 in cut Ivanhoe was made with improvements over time for the next eight years. The general-purpose Eagle and Golden Eagle roller mowers for fine turf appeared in 1935.

SUFFOLK

A Mr L J Tibbenham of Stowmarket established the Suffolk Iron Foundry in 1920 in order to manufacture castings for agricultural and electrical equipment. Within five years the Suffolk Iron Foundry was making various domestic items including lawn mowers, line markers and tennis posts, Fleetway laundry wringers, kitchen furniture and Clipper hand drills.

Further expansion came in 1930 with the introduction of Sifbronze oxy-acetylene welding equipment. Tank transporters and bomb trolleys were made at Stowmarket during the war years and,

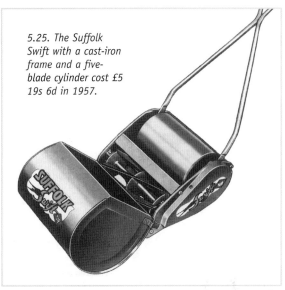

5.25. The Suffolk Swift with a cast-iron frame and a five-blade cylinder cost £5 19s 6d in 1957.

when the factory returned to peacetime, hydraulic pumps for farm tractors were also made.

The first Suffolk hand mowers were made in 1925 and the Suffolk Punch motor mower appeared in 1954. The 10 in cut Suffolk Swift and 12 in Super Swift roller mowers and the Super Clipper side-wheel mower were built at Stowmarket in the immediate post-war period. The Swift and Super Swift were still in production in 1958 when Suffolk Iron Foundry joined the Qualcast Group but the Super Clipper had given way to the 10 and 12 in cut Suffolk Viceroy and the 12 in Viceroy de luxe side-wheel mowers. The 10 in Viceroy cost £3 19s 6d and the 12 in Super Swift was 7 guineas (£7 7s 0d).

Suffolk Iron Foundry amalgamated with Atco in 1969 to form Suffolk Lawnmowers which had an annual output in excess of half a million hand mowers in the early 1970s. The Suffolk Lawnmowers catalogue for 1970 included the Swift, Super Swift and the Mk II Viceroy side-wheel mowers. The 10 and 12 in cut Mk II Viceroy had five-blade cutting cylinders running on sealed-for-life bearings, 7¼ in diameter steel wheels and tubular steel handlebars.

Some of the earlier Birmid Qualcast mower brand names, including the 14 and 16 in cut Atco Elite and the 14 and 16 in Suffolk Super Clipper side-wheel mowers together with the Webb Wasp and Witch roller mowers, were listed in the early 1980s Suffolk Lawnmowers catalogues.

5.24. The Suffolk Iron Foundry made the Super Clipper side-wheel mower in the early 1950s.

WEBB

Henry Webb & Co of Birmingham, manufacturers of hub brakes and spring forks for motor cycles, bicycle components and roller skates, made its first hand mowers in 1928. Within twenty years lawn mowers were the company's main product. The first model, the Webb de luxe roller mower, had an enclosed gear drive from the roller to the cutting cylinder and unlike any other mower at the time the side frames were made of pressed steel. Other makes had cast-iron side frames and Webb advertisements pointed out that its de luxe roller mower was half the weight of its competitors and less liable to breakage.

A new factory was built in 1929 and the 12 in Wasp with a six-blade cylinder which cost £1 19s 6d was added to the original 10 and 12 in cut de luxe models. The 10 in Whippet and high specification 12 and 14 in Windsor hand mowers appeared in the mid-1930s. Webb also made the chassis for motor mowers made by the Enfield Motor Cycle Co.

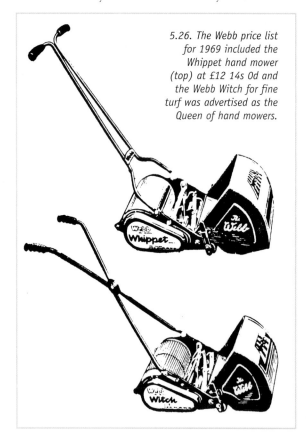

5.26. The Webb price list for 1969 included the Whippet hand mower (top) at £12 14s 0d and the Webb Witch for fine turf was advertised as the Queen of hand mowers.

5.27. The Webb Witch hand mower was still being made in the late 1970s.

Improved versions of the pre war Whippet, Wasp and Witch were introduced in 1947. The Whippet, described as a ladies' lightweight mower, weighed only 38 lb and was made until the late 1970s. The Wasp was a more robust machine with a 12 in, six-blade cylinder and the 12 in cut Witch had a split rear fluted roller to provide both differential and free wheel action and an eight-blade cylinder for fine turf which made approximately 60 cuts per yard.

H C Webb became part of the Wolseley Hughes group in 1963 but continued to trade as a separate company until 1973 when the two garden equipment manufacturers combined to form Wolseley Webb. The Whippet was discontinued before Wolseley Webb became part of the Qualcast group in 1984 but the Wasp and Witch hand mowers remained in production until 1988 when they cost £114.95 and £144.95 respectively

Motor Mowers

The first self-propelled lawn mower was made by James Sumner at Leyland in 1893. The steam-powered and pedestrian-controlled machine weighed 1½ tons and from cold it took about ten minutes for the boiler to build up a full head of steam. The Sumner family founded the Leyland Steam Motor Co in 1895 and two years later they were selling 25 and 30 in cut cylinder mowers for £60 and £90 respectively. Alexander Shanks made a ride-on steam mower in 1900 and Ransomes built the first petrol-engined ride-on motor mower in 1902.

Early motor mowers were large, cumbersome and better suited to large areas of grass such as sports grounds. The first petrol-engined motor mower suitable for domestic use was introduced by Atco in 1921, Ransomes built a mains-electric mower in 1926 and H C Webb made the first battery-electric mower in 1959. Power Specialities Ltd made the world's first rotary mower in Slough in 1933 and the Flymo rotary hover mower first appeared in 1964.

ARUNDEL COULTHARD

Arundel Coulthard & Co had started as a general engineering business based at Preston in 1815 but added hand-propelled and motor mowers to its product range during the depression years of the 1930s. It was making the 14 in cut Presto motor mower with a two-stroke petrol-engine in the late 1930s when it cost £18 18s 0d or on hire purchase terms with down payment £4 4s 0d followed by twelve monthly installments of £1 7s 3d. Mower production was resumed after World War II and Presto motor mowers were made until the late 1960s when Qualcast purchased Arundel & Coulthard and closed the business.

ATCO

Charles H Pugh opened a wholesale jewellery and ironmongers business in Rotherham, Yorkshire in 1865 but the jewellery was soon dropped in favour of engineering. After moving to Whitworth Works in Birmingham the company pioneered the development of the domestic motor mower. George Bull, the managing director of Charles H Pugh Ltd., used a donkey to pull his cylinder mower and when his donkey died he commissioned the works engineer to install an engine on his donkey mower. The project was a success and Charles H Pugh Ltd, the owner of the Atlas Chain Company, used the first two letters of Atlas and Company to form the Atco brand name for its motor mowers.

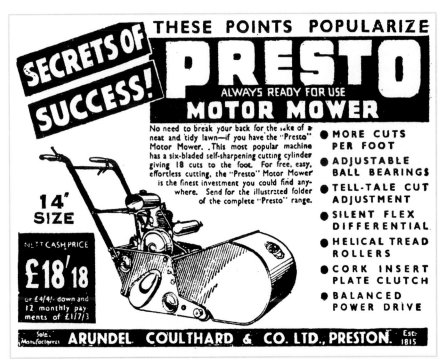

5.28. A 1937 advertisement for the Arundel Coulthard Presto motor mower. Qualcast acquired the company in the 1960s and promptly closed it down.

The first Atco motor mower, introduced in 1920

with a cast iron frame, a Villiers two-stroke engine and a 22 in wide cutting cylinder, cost £75. Within a couple of years Atco was also making 16 and 30 in cut motor mowers and had set up a network of depots around the country to deliver new machines and carry out any necessary repair work.

The 14 in cut Atco motor mowers made in the late 1940s and early 1950s had kickstart two-stroke 98 cc Villiers Midget engines with an aluminium rear roller and side plates. The Atco catalogue for 1959 included self-propelled 20 and 24 in side wheel motor mowers with two-stroke engines and the new

14 and 17 in cut four-stroke cylinder mowers with Atco-Villiers AV3 four-stroke engines. The specification included an improved kickstart and a simplified cutting height adjustment. The front roller was equipped with a scraper for greater mowing accuracy in wet conditions.

Atco joined the Qualcast Group in 1962 shortly after introducing a new Atco 14 in cut battery-electric mower and improved 12 and 14 in cut motor mowers. The new Atco 12 volt battery-electric mowers had a two-speed gearbox and a built-in battery charger which, according to press advertisements, were so much quieter and lighter than earlier models. The Atco catalogue for 1964 included 14 and 20 in cut petrol-engined motor mowers and battery-electric mowers with 17 and 20 in cutting cylinders. A 12 in cut motor mower appeared in 1965 followed by a new 14 in Atco battery mower in 1966. By the late 1960s there were twelve different Atco motor mowers from a 12 in domestic mower to a 34 in cut professional model.

The improved petrol-engined Atco 12 and 14 and Atco 17 and 20 in cut cylinder mowers together with the 12, 14 and 17 in cut battery-electric mowers were included in the 1969 Atco mower catalogue. The battery-electric cylinder mowers were popular in the late 1960s and early 1970s but sales gradually fell away and they went out of production in the mid-1970s.

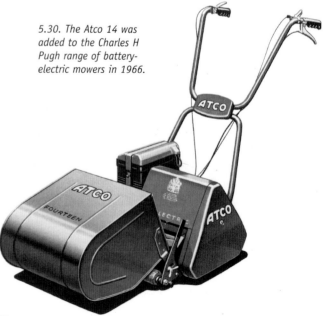

5.30. The Atco 14 was added to the Charles H Pugh range of battery-electric mowers in 1966.

5.29. Late 1940s Atco motor mowers had kickstart two-stroke engines.

There were eight sizes of Atco four-stroke motor mower in the early 1970s. The smallest 12 in cut mower cost £36 5s 0d and the widest 34 in cut 5 hp Villiers-engined Groundsman professional mower with a kickstarter or optional electric starter and trailing seat was £251. The 20 in side-wheel Toughcut cylinder mower introduced in 1971 for long grass had a 2 hp four-stroke engine, recoil starter, a multi-plate friction disc clutch, a five-blade cutting cylinder and a rear roller.

The more significant introductions in the 1974 Atco mower catalogue included 14, 17 and 20 in de luxe motor mowers and the new 28 and 34 in cut Mk II Groundsman with an 8 hp petrol engine.

The 12 to 20 in cut Atco Commodore B12, B14, B17 and B20 motor mowers, with the new 114 cc Atco aluminium engine, a single lever safety or deadman's handle on the handlebars and a plastic grass box replaced the Atco de luxe mowers in 1983. The mains-electric 12 in cut Mow & Rake with 75 ft of cable and built-in spring tines to remove thatch was added to the Atco lawn care range in the same year. An alternative four-stroke Mitsubishi engine with a recoil or electric starter was available for the 14, 17 and 20 in Commodore mowers from the late 1980s.

The Atco professional range of ride-on and pedestrian-controlled Royale and Groundsman mowers was also extended in 1983 with the addition of the ten-blade Club B17 and twelve-blade B20 de luxe motor mowers for fine turf. The 17 in cut Club B17 had an 88 cc Mitsubishi four-stroke engine with electronic ignition while a 206 cc Briggs & Stratton was used for the 20 in cut B 20 de luxe. Both models had a recoil starter and multi-plate disc clutch.

The Ensign B12 and B14 launched in 1989 for medium sized gardens were cheaper than the equivalent Atco Commodore mowers. The Ensigns had a 114 cc Atco engine with electronic ignition, a six-blade cylinder and a polypropylene grass box. The front roller could be replaced with side wheels

5.31. The Atco Toughcut mower was used to cut long grass.

5.32. The 1974 Atco 14 de luxe had a six-blade cylinder, a one-piece rear roller and a hardwood front roller.

COMMODORE B20   COMMODORE B14

5.33. Atco cylinder mowers in the early 1990s included three models of the Commodore with the choice of a 114 cc Suffolk or an 88 cc Mitsubishi four-stroke engine.

for cutting long or wet grass. The 17 in cut Ensign B17 and optional electric starting for the B12 and B14 were added in 1990.

Atco motor mowers in the mid-1990s included the Commodore and Ensign ranges, the 24, 28 and 30 in cut Royale with Briggs & Stratton power units and the Club B17 and 20 for fine turf.

The new mains-electric 14 in cut Atco Consort five-blade cylinder mower introduced in 1994 had 75 ft of cable, separate controls for the cutting cylinder and

driving roller, variable speed control and a polypropylene grass box.

BRITISH ANZANI

The British Anzani Engineering Co acquired the manufacturing rights for the 16 and 24 in cut Easimow tubular steel-framed ride-on motor mower with a ride-on roller seat in the mid-1950s. The 16 in cut Easimow with a 2½ hp JAP engine cost £96 and the 3½ hp Villiers-engined 24 in mower was £162.

5.34. The Heli-Strand flexible drive shaft for a hedge trimmer and other power tools could be used with the British Anzani Lawnrider. (Charlie Moore)

Ride-on mowers were rare animals when British Anzani advertisements explained that the Easimow 'mows as you ride as you roll'.

The more elegant and curvaceous British Anzani Lawnrider, initially with a two-stroke Villiers and later with a four-stroke Briggs & Stratton engine, replaced the Easimow in the late 1950s. The 24 in cut Mk III Lawnrider costing £125 0s 0d was the current model in 1964 when British Anzani also made the 14 in cut Wispa and Easimow and 24 in cut Powermow pedestrian-controlled cylinder mowers. By this time British Anzani motor mowers, which could be used with the Heli-Strand flexible drive shaft and range of tool heads, were made by the British Anzani Engineering Co at Aylesford near Maidenhead in Kent.

DENNIS

John and Herbert Dennis, based at Guilford, were making bicycles in 1895. Within ten years its products included cars, motor tricycles, box vans, buses and fire engines. Dennis Brothers introduced the first Dennis motor mower for use in public parks in 1922 and the company was still making a full range of vehicles in 1972 when it became part of the Hestair-Dennis Specialist Vehicles Group. The lawn mower side of the business was sold to Godstone Engineering in Surrey

and Dennis Godstone Engineering built Dennis professional mowers until the early 1980s.

Dennis Brothers was not involved with the domestic market in the early days; instead it concentrated on building 24 in cut professional cylinder mowers. Most Dennis mowers had an air-cooled single-cylinder 640 cc Blackburn four-stroke petrol engine and when the mower was not being used the hinged grass box could be turned upside down to keep out the rain. Three Dennis mowers were advertised in the mid-1930s when the 24, 30 and 36 in models cost £78, £95 and £110 respectively, while a roller seat was £6 10s 0d. Cash buyers were offered 5% discount and carriage was paid to the nearest railway station. The 1930s were busy times for Dennis Brothers and its annual output of about 1,000 mowers continued right up to the outbreak of war in 1939.

Similar Dennis mowers with 24, 30 or 36 in cutting cylinder and 7 hp four-stroke engines were listed in 1951 when a trailing seat was an optional extra. Improved 24, 30 and 36 in cut Dennis Premier Mk I motor mowers were made in the early 1960s. To illustrate their superior quality the 24 in Dennis mower cost £295 in 1964 compared with £104 for the 24 in Atco and £103 for the 24 in Webb. An optional flexible power take-off shaft for the Premier made it a more universal garden aid as it could be used with a chainsaw, hedge cutter, rotary pruning saw, hand-held rotary grass cutter and other attachments.

5.35. An optional flexible drive shaft for the Dennis Premier mower could be used with a hedge trimmer, a hand grinder, a rotary grass cutter or a small chainsaw.

5.36. Green's Zephyr motor mowers had ten-blade cutting cylinders for fine turf.

The 1970 Dennis mower price list included the 20 and 24 in cut Paragon mowers suitable for maintaining up to four acres of grass and the Premier with a 30 or 36 in wide cutting cylinder for estates and sports grounds. The Paragon had a 256 cc Villiers engine with independent car-type clutches to engage drive to the cylinder and rear roller, a six-blade cylinder, sealed bearings and a large polythene grass box.

The Premier was a heavy-duty mower with a four-stroke Dennis petrol engine. Sealed bearings meant there were no greasing points. Dennis mowers in the early 1980s included the 24 and 30 in Paragon with a six-blade or optional ten blade cylinder and 8 hp Kohler petrol engine and the 30 or 36 in cut Premier with a 12 hp Kohler petrol engine or optional 7.2 hp Kubota diesel.

GREEN

Thomas Green & Son of Leeds introduced its first 24 in cut motor mowers in 1903, shortly after the first Ransome motor mowers were made at Ipswich.

Greens were making 16 and 20 in cut light motor mowers in the late 1920s when, along with Ransomes and Shanks, it was the main contender in the motor mower market. The 16 in cut model had a 1¾ hp air-cooled two-stroke kickstart engine of Green's own design while a 2½ hp version of the engine was used for the 20 in cut mower. From 1938 Green 16 and 20 in cut mowers had a 147 cc Villiers or a 250 cc JAP four-stroke engine while a 98 cc Villiers Midget engine was used for the newer 14 in cut Green motor mower.

There were six models of Green motor mower in the late 1940s and early 1950s. The smallest 14 in cut Master had a 1 hp two-stroke Villiers engine while a 1.3 hp Villiers four-stroke was used for the 17 and 20 in cut mowers. Prices started at £45 7s 2d for the 14 in model and an optional hedge-trimming attachment was available for the three machines. Early 1950s Greens professional mowers included 24, 30 and 36 in cut machines with 3½ hp twin cylinder four-stroke engines, with an optional trailer seat available for £15 19s 1d.

The early 1960s Greens Zephyr mower for bowling greens and cricket pitches had a 1½ hp Villiers C-12 engine with a chain-and-sprocket drive cylinder and a three-section land roll. A dog clutch was used to disconnect drive to the six- or optional superfine ten-blade cylinder.

HAYTER

The first Hayter cylinder mower appeared in the late 1950s after Douglas Hayter discovered that one of his competitors, well known for its cylinder roller mowers, was moving into the rotary mower market. To meet this challenge he designed and built in a matter of weeks the self propelled Haytermower with interchangeable 30 in cylinder and rotary cutting units

The self propelled Ambassador launched in 1967 was the first purpose-built Hayter cylinder mower. There were two models, the 16 in Ambassador which had a 119 cc BSA engine and the 20 in mower which had a 2¼ hp Briggs & Stratton power unit. A professional version of the 20 in Ambassador with a Villiers four stroke engine and ten-blade cylinder for fine turf was added in 1969.

The improved Ambassador 2 with a five-blade, 20 in cylinder and 3 hp Briggs & Stratton engine was announced in 1973 followed by the 10-blade Ambassador Super 2 in 1979.

The Hayter 30 Condor introduced in the late 1960s with a 6 hp MAG engine and three-speed gearbox was designed for parks and sports grounds. Like the earlier Haytermower its 30 in cutting cylinder could be replaced with a 30 in twin-bladed rotary mower deck. The Hayter Senator 30, which replaced the Condor in 1980, had an 8 hp Kohler engine and a hydrostatic drive unit with infinitely variable speeds of up to 5 mph in both directions. The number of cuts per yard by the cylinder depended on the forward speed and a differential unit was built into the split rear drive rollers. Two versions of the 20 in cut Ambassador 3 with five- and ten-blade cylinders and a Briggs & Stratton Quantum engine replaced the earlier model in 1992. The Senator 30 and both models of the Ambassador 3 were still being made in 1994.

5.37. The Haytermower could be used with a 30 in cut cylinder mower or rotary mower deck.

5.38. Introduced in 1967, the Ambassador was the first Hayter cylinder mower for domestic lawns.

J P

Jerram & Pearson established J P Engineering at Leicester in 1919. The first J P motor mowers had a water-cooled engine which could be removed from the mower and used to drive other equipment. J P motor mowers were heavy machines with cast-iron side plates that were justifiably advertised as the Rolls Royce of Lawn Mowers.

5.39. Old Tom Noakes never found a better mower than his JP mower!

The 24 in cut J P motor mower introduced in 1927 with an own-make water hopper-cooled four-stroke engine and a large plywood-sided grass box cost £68 5s 0d which was far too expensive for the domestic gardener. The mower was chain driven through a two-speed gearbox, the low speed being used for fine turf mainly on sports grounds and the high speed used for cutting areas of recreational grass and for transport.

A heavy-duty 16 in cut J P mower introduced in 1929 was followed later by the 24 in cut J P 24. A four-stroke Blackburn engine replaced the earlier J P power unit in 1938 but production of J P 16 and J P 24 mowers ceased in 1939. With the war at an end J P Engineering made J P Super Simplex motor mowers. The range included 14 and 16 in cutting cylinders with a 1 hp two-stroke engine and ½ hp mains-electric cylinder mowers with a 12, 14 or 16 in wide cutting cylinder.

MOWER PUSHERS

William Edgecumbe Rendle obtained a patent in 1914 for a motor-driven machine to propel lawn mowers and other slow-moving machines. The first Rendle Mower Pusher was made by W Edgecumbe Rendle & Co of London in 1920. It had a 350 cc Villiers two-stroke engine on a steel frame with a countershaft and chain reduction drives to two drive rolls. The mower pusher, which could be used with almost every make of roller lawn mower, was attached to the back of the lawn mower with a metal pin. It had a trailing seat and was supplied with extensions for the handlebars.

Other companies introduced motor conversion kits for hand-propelled lawn mowers in the early 1920s. The MP Co in London's Oxford Street, which was connected with Ransomes, Sims & Jefferies, made the MP Mower Pusher. It was similar to the Rendle but had a seat over the engine. There were three different sizes and Ransomes suggested that the biggest MP Mower Pusher, with a fan-cooled two-stroke 269 cc Villiers engine, could be used to motorise donkey and pony mowers.

The Simplex mower pusher, made by Small Engines Ltd in Birmingham in the early 1920s, had an expanding frame which could be adjusted to fit most makes of 12 to 18 in cut hand mowers. Complete with a clutch and starting handle, the Simplex cost £18 10s 0d.

5.40. A sales leaflet suggested that the Trusty Mowmotor would put an end to the labour and sweat of pushing a hand mower.

The Trusty Mowmotor made by Tractors (London) Ltd in the mid-1950s was advertised as a sturdy, dependable little job at a reasonable cost which would fit any ordinary mower and do two or three hours work on a pint of fuel. It had a ⅛ hp two stroke engine and the weight of the Mowmotor held two small-diameter driving rollers firmly in contact with the rear roller on the mower. The rollers were chain driven from the engine and a lever on the handlebars was used to raise and lower the engine unit to start or stop forward travel.

QUALCAST

The first Qualcast motor mower was made in 1948. Designated the Qualcast 16, it had a 98 cc Villiers two stroke engine, a 16 in cutting cylinder and cost £47 18s 7d complete with a grass box. The 16 in cut, engine-driven Commando side-wheel mower for long grass was announced in 1953 and the 16 in Royal Blade motor mower priced at £48 7s 4d was added to the Qualcast range in 1954.

The Powered Panther roller mower was a motorised version of the hand-propelled Qualcast Panther roller first made in 1932. The Powered Panther, which cost £24 19s 6d when it was introduced 1954, was not self propelled, but publicity material pointed out that the power-driven cutting cylinder lightened the gardener's load. The 34 cc JAP two stroke engine used about one pint of fuel per hour and a foot pedal was used to engage drive to the cutting cylinder which was automatically disengaged when the operator took his or her hands off the handlebars.

The Qualcast Royal Blade de luxe, the Powered Panther and the Commando were still being made in 1957 but the Commando was discontinued in 1960. Qualcast and Suffolk Iron Foundry Ltd retained its own models of motor mower when the Stowmarket company joined the Qualcast Group in 1958. Early co-operation was evident when the Qualcast Commodore was launched in 1959 and, unlike the two stroke engined Qualcast Royal Blade de luxe, the new 14 in Commodore with dual clutch control and recoil starter had a Qualcast Suffolk 75 G four stroke engine.

5.41. The Qualcast 16 motor mower was introduced in 1948.

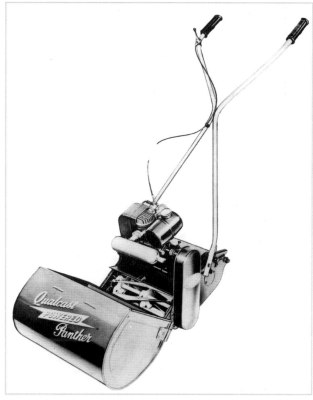

5.42. A foot pedal was used to disengage the drive to the Qualcast Powered Panther cutting cylinder.

5.43 The Qualcast Royal Blade de luxe had a recoil starter and dual clutch control.

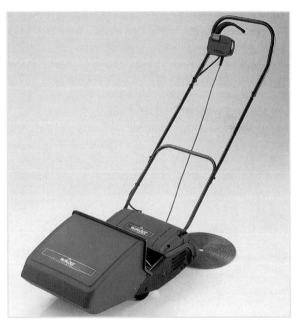

5.45. The mains-electric Qualcast Concorde was introduced in 1971.

The range of Qualcast and Suffolk motor mowers was gradually reduced after Atco joined the Qualcast Group in 1962. The 12 in cut Qualcast Super Panther Electric, introduced in the same year, was a self propelled mains-electric mower with a 0.4 hp 220/250 volt motor, 75 ft of cable, a dual-speed drive

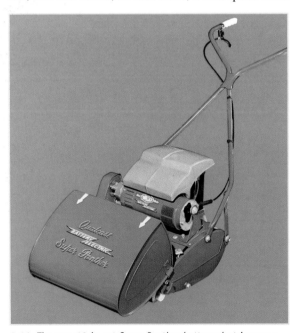

5.44. The new 12 in cut Super Panther battery-electric mower was launched in 1967.

for cutting in confined spaces and a swinging arm on the handlebars to keep the electric cable away from the cutting cylinder.

The 14 in cut Super Panther Electric added in 1965 cost £29 19s 6d and its three speed drive was said to allow the user to walk at a pace to suit their available energy and the length of the grass being cut. The battery-electric 12 in cut Super Panther was announced for the 1967 season when, complete with a built in 12 volt battery charger, it cost £29 19s 6d. It was an immediate success and even though about 10,000 were produced in the first year demand for the battery-electric Super Panther outstripped supply, so twice as many were made in 1968 and it remained in production for another eleven years.

The 12 in cut Qualcast Concorde electric cylinder mower was launched in 1971 when complete with 50 ft of cable it cost £15.75. The Concorde was a huge success and almost half a million were sold in the first two years. This led to the introduction of the 14 in cut Astronaut with 75 ft of cable in 1974 but it was not so popular and was discontinued in 1979.

The Qualcast Concorde has been made in huge numbers in various forms over the years including the 12 in Concorde E30 for short or long grass introduced in 1980, the Concorde RE30 DL with a rear grass box and the 14 in cut Concorde RE35 DL. The Concorde RE30 S and RE35 S cylinder mowers added in 1983 were supplied with lawn raking kits used to convert the mower to a powered rake.

The Concorde was still selling in big numbers in the mid-1990s when the range included the 12 and 14 in cut E30 and E35 with front grass box and the 10, 12 and 14 in XR models with a rear grass box.

RANSOMES

Ransomes' first petrol-engined cylinder mower (page 205) was made in 1902 but by the mid-1930s there were at least nine different models of Ransomes motor mower. The smallest 14 in cut Midget with a 1 hp air-cooled two-stroke engine cost £22 17s 6d and the largest 11 hp ride on mower with a 42 in cutting cylinder and weighing 22 cwt was £325, less five per cent discount for a cash payment. The 42 in cut mower had a water-cooled four-cylinder engine, band clutch, a forward and reverse gearbox and separate chain drives to the cutting cylinder and drive roller.

The Ransomes Electra, the first mains-electric lawn mower to be sold in any quantity appeared in 1926. The 14 in cut model with a 1½ hp motor weighing 1½ cwt was said to cut half an acre in an hour and consume half a unit of electricity. The Electra was also made with 16 and 20 in wide cutting cylinders while a 30 in model was made to order. A 1933 price list shows the 14 in mower costing £27 10s 0d and the 20 in model £57 10s 0d. The lightweight 20 in cut Bowlic mains-electric cylinder mower was made for bowling greens where, as there was no risk of petrol or oil spillage, it was popular with green keepers and was made for many years. A Ransomes catalogue noted that the main advantages of electric mowers were their cleanliness, silent operation, easy manipulation and economic running.

5.46. This early 1950s Ransomes 40 in cut ride-on mower had a 10 hp four-cylinder petrol engine and weighed one ton.

5.47. The Bowlic mains-electric mower with a centrifugal clutch and a ten-blade cuttting cylinder for bowling greens and other areas of fine turf cost £104 17s 10 in 1956.

Manufacture of mains-electric and petrol-engined motor mowers ceased during the war years. A few gang mowers were made for cutting airfields but most of the factory was used for armament production. With hostilities at an end Ransomes was able to make limited numbers of lawn mowers including the 20 and 24 in Electra and 16, 20 and 24 in cut motor mowers.

The 18 in cut mains-electric Cellec mower for bowling greens which appeared in the late 1940s had a ½ hp 240 volt motor which used half a unit of electricity to drive its eight- or ten-blade cylinder in one hour. The pre-war ½ hp mains-electric Bowlic with a ten-blade cylinder and centrifugal clutch was still being made in the mid-1950s when with 75 yards of cable it cost £104 17s 10d. The ½ hp Mk 7 and 1hp Mk 7A Cellec mowers were current in 1960 when improvements included a chain-driven rear roller and a clutch to disengage the cutting cylinder so that the Cellec could also be used as a roller.

Ten two and four stroke motor mowers listed in Ransomes motor mowers catalogue for 1952 included the 12 and 14 in cut self propelled Mk 6 Minor and the 18 in hand propelled Auto Certes with 98 cc two stroke Villiers engines and a centrifugal clutch. The Auto Certes used for bowling greens and croquet lawns had a

5.48. The Mk 7 Minor motor mower launched in 1956 had a separate clutch for the driving roller.

ten-blade cylinder which made 15 cuts per in at 3 mph. The 147 cc Villiers two stroke engined Ransomes 16 with a centrifugal clutch had separate dog clutches to engage the drive to the two-piece land roll and to the 16 in wide cutting cylinder. The Ransomes 16 with a seven-blade cylinder was recommended for private lawns and tennis courts.

5.49. Ransomes claimed that the Meteor had no equal for keeping lawns of limited size in a state of constant perfection.

5.50. The Ransomes Antelope, which replaced the Gazelle in 1956, was made for thirty-seven years.

Ransomes made 16, 20, 24 and 30 in four-stroke motor mowers in the early 1950s. Ransomes 248 cc engines were used for the 16 and 20 in cut mowers; a 348 cc unit powered the 24 in mower and the 30 in model had a 600 cc Ransomes engine. A centrifugal clutch engaged the main drive, and separate dog clutches were used to control the cutting cylinder and rear drive roller.

Ransomes' 40 in cut ride-on motor mower and roller, which weighed one ton, had a four-cylinder water-cooled Ford 10 hp four-stroke engine, a six-blade cutting cylinder and a centrifugal clutch used in conjunction with a lever-operated epicyclic forward and reverse gearbox. The engine with a 6 volt coil ignition system and force feed wet-sump lubrication system used about five pints of petrol per hour.

Introduced in 1951, the 18 in cut Gazelle side-wheel mower for cutting rough grass had a 98 cc Villiers two stroke engine and light alloy die-cast frame. In 1957 the Gazelle was replaced by the 20 in cut Antelope side-wheel mower which had a 119 cc BSA engine and a five-blade cylinder. The Antelope was improved several times during its long production run which came to an end in 1993.

The Ransomes Verge Cutter with 30 in cutting cylinder for rough grass was introduced in the early 1950s. It had a Ransomes four-stroke 348 cc air-cooled engine, a centrifugal clutch with separate dog clutches for the cutting cylinder and large-diameter, low-pressure pneumatic-tyred wheels. The driving wheels on a new model of the Verge Cutter introduced in 1954 were at the back and within the width of the 30 in cutting cylinder so that it could cut up to the edge of roadside verges. A pair of castor wheels supported the front of the machine and a trailing seat was offered as an optional extra.

In 1956, the Ransomes motor mower catalogue explained that the Minor Mk 6, the new Minor Mk 7 and new Meteor two-stroke Villiers-engined mowers had no equal for keeping lawns of a limited size in a constant state of perfection. Ransomes four-stroke mowers were described as quality-built machines providing the most efficient means of keeping large lawns, parks, estates and playing fields in first-class condition.

The catalogue also included the Auto-Certes, Bowlic and Overgreen for golf courses, the Gazelle and Verge Cutter for longer grass, the heavyweight 40 in cut ride-on mower and roller and the Aero-Main for turf care. Unlike the 12 and 14 in cut Mk 6 Minor with its car-type starting handle, the new six-bladed 14 or 18 in cut Mk 7 Minor cylinder mower had a kickstart 98 cc Villiers engine, centrifugal clutch and a separate clutch for the roller drive.

The new 20 and 24 in cut Meteor motor mowers had a 147 cc Villiers two-stroke engine with a geared-up kickstart pedal, centrifugal main clutch and separate clutches to disengage drive to the cylinder and roller drive. A cable-operated brake for the rear roll and transporting carrier wheels were optional extras.

Ransomes' four-stroke mowers included the 20 in cut model with a 248 cc engine, the 24 in mower with a 348 cc engine and the 600 cc 30 and 36 in cut machines. Ransomes four-stroke side-valve dry sump engines with pressure and scavenging oil pumps, centrifugal main clutches and secondary clutches for the cylinders and roller drive were standard across the range.

Also included in the 1956 price list were the Ransomes 40 in cut mower and roller with its 10 hp Ford engine and a work rate of 1½ acres an hour; the 348 cc engined Verge Cutter and Overgreen mowers, and the Aero-Trac power unit.

Ransomes' sales literature explained that the Mercury,

5.51. The 348 cc Aero-Trac power unit could be used with the Sisis Aero-main turf aerator, 36 in cut Ransomes Aero-cutter cylinder mower and other turf care equipment.

Marquis and Sprite motor mowers introduced in the late 1950s were designed for the discriminating gardener. The 14 in cut Sprite with a 34 cc JAP two stroke engine and fingertip dual controls on the handlebars was claimed to cut 500 square yards of grass for less than 1d. The last Sprite mowers were made in 1964.

The 16 in cut Mercury with a 75 cc Villiers four stroke engine was made between 1958 and 1965. The 18 in and 20 in cut Ransomes Marquis with a 119 cc BSA four-stroke engine, centrifugal main clutch and separate clutch to control the drive to the rear roller was popular with professional users and private gardeners. A de luxe version appeared in 1964 and the Marquis motor mower was still being made in 2001.

The Ransomes Fourteen, also announced in 1964, was advertised as a quiet-running quality mower, packed with power and made to last. Made for six years, it was described in sales literature as a slim-line machine light manoeuvrable enough for any member of the family to use. The petrol version with a BSA 65 cc four stroke engine, later replaced by a 65 cc Villiers, cost £42 10s 0d although the 12 volt battery electric model was £52 10s 0d.

Ransomes' mid-1960s catalogue included the 20 in Antelope side-wheel mower, the Fourteen and the Marquis motor mowers. To meet a growing demand for

5.52. The Ransomes Fourteen was made between 1964 and 1970.

battery-electric mowers a 12 volt battery-electric version of the Ransomes Fourteen motor mower was made from 1964 to 1969. Ransomes withdrew from the domestic mower market in the mid-1970s to concentrate on professional mowers and grass care equipment including the Twenty-four, Marquis, Matador and Mastiff. The Twenty-four was discontinued in 1981 but the other motor mowers were still being made in 1995 - some 160 years after J R and A Ransome were first granted a licence to manufacture Edward Budding's gear-drive mower

ROYAL ENFIELD

The Enfield Cycle Co. was one of a number of companies to turn to lawn mower manufacture during the depression years and four models of Royal Enfield motor mower were made at Redditch in the mid-1930s. The company motto 'Made like a Gun' was an apt

FOOT STARTER
-NO STOOPING
ALL-GEAR DRIVE
-NO ADJUSTMENT

Write for Catalogue—
"Lovely Lawns, less labour," which shows our four models.

Prices from
£18.10.
or 9/- weekly

Royal Enfield
MOTOR MOWERS

● Efficient cooling of engine by Blower.
● High-speed cutting cylinder.
● 75 cuts per yard — for fine cutting.

Touch the starter with your foot and off you go for a trouble-free stroll, with never a hitch to check the good work until your lawn is like a carpet. It's truly a "walk-over" with a Royal Enfield.

THE ENFIELD CYCLE CO., LTD., Dept. M.G., REDDITCH
London Showrooms: 48 Holborn Viaduct, E.C.1

5.53. A 1938 advertisement for Royal Enfield motor mowers.

description for the 22 in cut Royal Enfield motor mower which cost £45 in 1935. The grass box had a pair of wheels and handlebars to make it easier to empty.

Four Royal Enfield mowers with two-stroke, kickstart, fan-cooled engines and gear drives to the roller and cutting cylinder were listed in the Royal Enfield 'Lovely lawns, less labour' catalogue for 1938. Prices started at £18 10s 0d or on easy terms of 9s 0d weekly. Post-1945 Royal Enfield motor mowers included a 14 in cut model with a 1 hp two-stroke engine which cost £45 5s 0d in the early 1950s.

SHANKS

The Firefly and Dragon motor mowers with four-stroke engines were made by Alexander Shanks in Arbroath in the early 1950s. The 16 and 20 in cut Firefly was for domestic lawns. A steering trailer seat was available at extra cost for the Dragon professional motor mower with a 30 or 36 in cutting cylinder. The makers estimated that the 16 in cut Firefly would cut about a quarter of an acre in an hour while the 36 in cut Dragon would trim 1½ acres in the same time. Shanks also made the 12 in cut mains-electric Pilot cylinder mower which cost £23 16s 3d in 1951.

SUFFOLK

The Suffolk Iron Foundry (1920) Ltd introduced the famous Suffolk Punch motor mower in 1954. Within three years the Suffolk motor mower catalogue included the Standard, Super and Professional models of the Punch, the 12 in cut Auto Swift and the Suffolk Pony. The Suffolk Iron Foundry became part of the Qualcast Group in 1958.

The Suffolk Pony had a 50 cc Suffolk two stroke engine and a dual-drive arrangement, which enabled the user to disengage the drive to the roller and push the machine into awkward corners and other confined spaces. The Punch had a recoil starter and a centrifugal clutch. The 14 in cylinder was mounted on self-aligning ball bearings; dual-drive, which enabled the user to disconnect drive to the roller, was added in 1956.

Sales literature explained that the four stroke engine on the 14 in cut Punch only used quarter of a pint of fuel to cut 500 square yards of grass in less than twenty minutes. The 17 in Super Punch with dual-drive introduced in 1956 was followed by the Super Punch Professional with a ten-blade 17 in wide cylinder for fine turf.

The Squire and the Squire Corporation four stroke engined side-wheel mowers with a centrifugal clutch and rubber-tyred wheels completed the 1956 range of Suffolk motor mowers. The Squire Corporation with Suffolk 75G four-stroke engine, heavy-duty bottom blade and a safety overload clutch to protect the drive to the 19 in cut cylinder became the Suffolk Corporation in 1960. The 12 in cut Suffolk Colt motor mower with a four stroke engine, automatic clutch and dual drive replaced the Pony in the same year. The new Mk II Corporation side-wheel mower and various models of the Punch and Colt were included in the 1965 Suffolk Lawn Mower catalogue.

The 12 in cut Super Colt introduced in 1968 was described as streamlined in construction and its many special features included an automatic clutch, dual drive operated from the handlebars, a unique bottom blade assembly and specially hardened cutting cylinder blades. The 1969 Qualcast catalogue included the Suffolk Colt and four models of Suffolk Punch. The Mk II Corporation was £29 19s 6d and the Squire had given way to the self propelled side-wheel Suffolk 16 with a Suffolk 75G four stroke engine, centrifugal clutch and five-blade cylinder.

Suffolk Iron Foundry also added the new Microset cylinder adjustment to the 14 in Punch and announced the Carry Mow mower transporter in the same year. The Carry Mow was a wheeled cradle placed under the main roller making it easier to push the Suffolk Punch and other mowers up and down steps or across gravel paths. The Carry Mow cost £3 9s 6d and with additional fittings priced at £2 4s 0d it could be used as a sack holder or a household bin carrier.

In mid-1970s the only petrol-engined Suffolk mowers made were the 12 in cut Super Colt, the 14 in Super Punch, the 17 in Super Punch, the 17 in Super Punch Professional and the Mk II Corporation.

The 1979 Birmid Qualcast catalogue listed the Mk II Suffolk Corporation self-propelled 19 in cut mower with a 98 cc Suffolk four-stroke engine and five-blade cylinder for rough grass. Also included were the Super Colt, 14 in

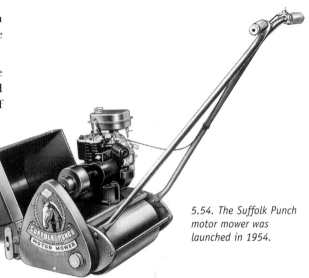

5.54. The Suffolk Punch motor mower was launched in 1954.

and 17 in Super Punch and the Super Punch Professional motor mowers.

The 12 in cut two-speed Punch EP30 and 14 in three-speed Punch EP35 mains-electric mowers with front and rear rollers designed to give a striped lawn were introduced in 1980. Both were double insulated and had an automatic clutch, adjustable pram-type handles and a grass box. A 98 cc four-stroke Suffolk engine was used for the new Suffolk Punch 30, 35 and 43 mowers

5.55. Introduced in 1960, the Suffolk Iron Foundry Carry-Mow was the easy way to move a motor mower up and down steps and along gravel paths.

5.56. An early 1990s 17 in cut Suffolk Punch with a 148 cc four-stroke petrol engine.

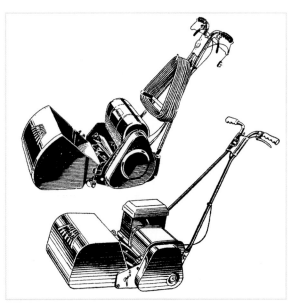

5.57. Electric mowers made by H C Webb in the late 1960s included a six-blade mains-electric model which cost 1d an hour to run and a battery-electric mower with a built-in charger.

with 12, 14 and 17 in wide cylinders and a front polypropylene grass box. The Suffolk Punch Professional 43 had a ten-blade cylinder, machined steel front roller and smooth rear roller for cutting fine turf.

The mid-1980s Punch range included the hand- and self-propelled mains-electric Punch Classic 30 and 35 and the Suffolk Punch 30, 35 and 43 with more powerful 114 cc Suffolk petrol engines.

Suffolk Punch cylinder mowers made by Atco-Qualcast at Stowmarket in the mid-1990s included 14 and 17 in cut motor mowers and a 14 in battery-electric model. They both had the QX Quick Exchange system with an easily removed cylinder cassette interchangeable with a lawn scarifier for power-raking.

WEBB

H C Webb of Birmingham, which introduced the first mains-electric cylinder mower in 1947, made the 12 in cut Wasp and 14 in cut Webb Model A and B mains-electric mowers in the early 1950s. Webbs created a sensation at the 1959 Chelsea Flower Show when demonstrating a 12 volt battery-electric mower with the added gimmick of radio control. By the mid-1960s the company was manufacturing two-speed 12 and 14 in cut battery-electric mowers with a built-in 1½ amp charger

and a de luxe 14 in model with a 3 amp charger.

The 1969 Webb catalogue included 12, 14 and 18 in cut mains-electric and two-speed battery-electric mowers which cost less than a penny an hour to run. Low-voltage mains-electric cylinder mowers with a built-in transformer to reduce the power to 110 volts were introduced in 1973 and both mains-electric and battery-electric versions of the Webb Wizard mower were added in the late 1970s.

The Webb 12 and 16 in cut motor mowers introduced in 1958 were described as very fine and sophisticated machines. The engine and clutch on the 16 in model could be removed and used with a flexible shaft to drive other garden equipment including a hedge trimmer and hand tiller. The 14 and 18 in cut Webb petrol mowers with Briggs & Stratton 2 hp engines and dual drive were introduced in 1961.

In the 1950s cylinder mower attachments were made for the Wolseley Merry Tiller and other garden tractors. The Merry Lawn Mower was attached to the Merry Tiller rotor shaft housing and forward travel was controlled with the cultivating rotor clutch. A vee-belt driven 16 in cut Webb lawn mower with a separate clutch made it possible to push the Merry Lawn mower round small flower borders and into awkward corners.

In 1963 H C Webb joined the Wolseley Hughes group, took control of the lawn mower side of the business and carried on trading as a separate concern under the Webb name. A 24 in cut ride on mower with a trailed bucket seat on a roller introduced in 1963 was described by sales literature as a fast, wide and handsome machine. It was also suggested that new owners could leap aboard and ride ahead of their cutting problems, as the new mower would care for the lawn and its owner.

The close-coupled seat was claimed to give the machine a turning circle that made a London taxi look like the Queen Mary. The 24 in ride on mower, which cost £32, had ribbed rubber-covered rollers, a lever-operated clutch, double vee belt transmission and a foot brake.

The company changed its name to Webb Lawnmowers Ltd in the late 1960s. Within ten years the major part of the Wolseley factory was devoted to the production of Webb lawn mowers. The 1969 Webb mower range included 24 in ride-on and pedestrian-controlled mowers and new 14 or 18 in Webb motor mowers for medium and large sized lawns with a 2 hp Briggs & Stratton four-stroke engine and vee-belt drive to the five-blade cutting cylinder.

The 24 in cut mower had a four-stroke Villiers engine with optional Dynastart – a combined generator and starter motor – and a twin vee-belt drive to the cutting cylinder and rubber-covered rear roller.

Wolseley Webb introduced the Webb AB series motor mowers in the early 1970s. The 18 and 24 in cut models were pedestrian-controlled and the 24 in cut mower was equipped with a riding seat. The Webb AB series mowers became the Webb Wizard motor mowers in the late 1970s when the Wizard range included 12 and 14 in cut mains-electric and 12 in battery-electric models. The hand-propelled 12 in cut mains-electric Wizard cost £33.73 and the most expensive 12 in battery mower £82.22.

A new 21 in cut Webb motor mower, launched in 1981, had a 3 hp Briggs & Stratton engine and a six-blade cylinder running on self aligning ball bearings. New models in 1982 included the 14 and 18 in cut Webb mowers with 2 hp Briggs & Stratton engines and a 5 hp pedestrian-controlled or ride on 24 in cut mower.

Wolseley Webb joined Suffolk Lawnmowers in 1984 and, except for the 24 in Webb motor mower, it was all change in 1985 when new designs brought Webb petrol

5.58. The Merry Lawn Mower could be attached to the Merry Tiller without tools in a matter of minutes.

models into line with the Suffolk Punch. Webb motor mowers have not been made since 1988 except for the brief appearance of the Webb Diplomat between 1990 and 1991.

5.59. Early 1980s Wolseley Webb lawn mowers had Briggs & Stratton engines.

Rotary Mowers

Power Specialities at Maidenhead made the world's first rotary mower in 1933. The company was bought by J E Shay in 1952 which became part of Wolseley Engineering in the early 1960s.

Hayters of Spellbrook, Herts, a name synonymous with rotary mowers, entered the domestic lawn mower market in 1947 with the launch of the 24 in Motor Scythe. Vivian Loyd at Camberley in Surrey introduced the Loyd Motor Sickle in the late 1940s. There were no guards over the vee belt drive and high-speed cutting rotor and both machines tended to be rather dangerous with stones and debris liable to fly in all directions.

The growing popularity of the domestic rotary lawn mower in the early 1950s coincided with the development of efficient, lightweight vertical crankshaft engines. Another milestone was reached in 1964 with the introduction of the heavily patented petrol-engined Flymo rotary hover mower. Britain's gardeners spent around £140 million a year on new lawn mowers in the mid-1980s. About 20 per cent had a petrol engine and the rest were driven by mains- or battery-powered electric motors.

ALLEN

The hand-propelled Allen Junior Rotary Sickle introduced in 1959 was the first rotary mower made by John Allen & Sons (Oxford) Ltd. The Allen Rotary Sickle, introduced in 1960, was a two-speed, self propelled version of the Junior Rotary sickle. An advertisement explained that with its four-blade rotor running at 2,600 rpm the Rotary Sickle made 10,400 cuts per minute and would make short work of cutting long grass and overgrown vegetation.

The self propelled heavy-duty 4½ bhp Allen Champion Three launched in 1962 had a Briggs & Stratton engine and a vee-belt drive to the 26 in rotor. The wheels, with ratchets in both hubs, were driven by a worm drive reduction box, a three forward and one reverse Albion gearbox and roller chain final drive. The Champion Three, which cost £140, was described as the de luxe model in the Allen rotary mower range.

Allen introduced the ride-on Motostandard garden tractor (page 25) at the 1961 Smithfield Show. The Roper was launched in 1973 and various models of Motostandard/Gutbrod dual-purpose garden tractors with rotary mower decks were sold in the UK until 1987.

John Allen & Sons introduced new models of rotary mower at frequent intervals, advertising the self-propelled Allen Cavalier, Challenger and Champion with Briggs & Stratton four stroke engines in 1963. The single-speed 20 in cut Cavalier for the domestic garden had a rear grass collector which could be replaced with a deflector flap for dealing with knee-high grass and weeds.

The 22 in cut Challenger with a direct vee-belt drive to the four-blade rotor and a geared reduction unit combined with a single belt drive to the rear wheels was recommended for cutting long grass in paddocks and orchards. The heavy-duty 6 hp Allen Champion also had a direct vee-belt drive to an aluminum rotor with

5.60. The Allen Lawnman had a 3½ hp Briggs & Stratton engine, a polypropylene grass collector and folding handlebars.

four replaceable blades. The rear wheels were driven by a worm reduction unit, three-speed Albion gearbox and roller chain final drive. Cutting height could be varied from ¼ in to 2¼ in with a micrometer-type adjuster.

Allen acquired Mayfield Engineering in the mid-1960s after which various models of Allen Mayfield garden tractor with rotary mower attachments (page 59) were made until 1984. The Mk 21 with a 21 in cut rotary mower and 24 in cylinder mower attachments was the only Allen Mayfield model made solely for cutting grass. Other Allen Mayfield attachments could not be used with the Mk 21 and its rotary and cylinder mowers did not fit other Allen Mayfield machines

After road testing a prototype 26 in cut ride-on Allen Mayfield Merlin (page 252) from Littlehampton to Oxford, Allen Power Equipment manufactured 109 Merlin ride-on rotary mowers between 1966 and 1968. In the late 1960s Allen also made its own range of domestic rotary mowers, including the self-propelled Champion and Challenger and the hand-propelled 18 in cut Allen Lawnman.

Allen domestic and professional rotary mowers in the early 1970s included the heavy-duty Champion Three with three forward speeds and one in reverse, the 22 in cut Challenger Mk 4 and three versions of the 19 in cut Allen 19 rotary mower. The Allen 19 Export had a 2½ hp two-stroke engine with a recoil starter. A 3½ hp four-stroke engine was used for the Allen 19 Special and a 4 hp two-stroke engine powered the Allen 19 Professional. The Allen Challenger Mk 5 Commercial with a 4 hp Briggs & Stratton engine, made in the early 1970s, was replaced in 1974 by the Challenger Mk 6 with a 4 hp Villiers Vertex engine.

When patents for the Flymo air-cushioned mower expired in the mid-1980s, Allen Power Equipment was one of several companies to introduce air cushion or hover mowers. The 1984 Allen Crown range of air cushion mowers included a 15 in cut mains electric model, 15, 18 and 21 in cut hover mowers with two-stroke Tecumseh engines and two four-stroke 4 hp Briggs & Stratton-engined models with 18 and 21 in cutting rotors. The smaller lightweight mowers were equipped with a carrying handle while the 21 in cut professional models had a pair of detachable transport wheels.

5.61. The fibre-glass bonnet gave the Allen Challenger a streamlined appearance.

5.62. The 18 in cut Allen E18 mains-electric air cushion mower had 15 m of cable and a dead man's handle safety switch.

ARWELL

Made in the late 1940s by the R Wells Agricultural Group in London, the hand-propelled Arwell was a two-in-one rotary scythe and vertical hoe for small plots of land. It had a 0.8 hp four-stroke JAP engine and two axles so that it could be used with one or two pneumatic tyred wheels.

The Arwell was used with two wheels when cutting fine and coarse grass, hay, kale and potato haulm with the rotary scythe attachment. The vee-belt driven 18 in wide rotor had three double-edged blades and by using the belt in a crossed or uncrossed position the rotor could cut in a clockwise or anti-clockwise direction. For hoeing, the Arwell was used with a single wheel and its three hoe blades on a horizontal rotary disc could be used to work in vegetable crops planted in rows up to 10 in apart.

ATCO

The four-stroke Atco 18 and 21 in cut rotary mowers introduced in the late 1950s were superseded in 1962 by an improved model with the Atco single lever Heightomatic height adjustment which altered the setting of all four wheels at the same time.

5.64. The Atco B18 mulcher mower.

Launched in 1969 at £58, the new Atco 18 in self-propelled rotary mower with a four-stroke engine and a rear grass collector was still being made in the mid-1970s. The cutting rotor was mounted on the engine crankshaft and the rear roller was vee-belt driven with a tensioner pulley to engage drive to a worm reduction unit and chain final drive. The Atco B18 mulcher mower with a 4 hp Briggs & Stratton engine was introduced in 1980. It retained the cut grass under the specially shaped rotor hood while the blades, which made 7,200 cuts per minute, chopped the trimmings to a fine mulch before discharging them from the rear. The Atco B45 mulcher mower, also with an 18 in rotor, replaced the B18 in 1981.

Three models of Airborne air cushion rotary mowers with two-stroke Aspera engines to drive the cutting rotor and fan appeared in 1980. The 16 in cut Atco Airborne B40 and 19 in cut B48 mowers had 98 cc engines and a 125 cc power unit was used for the 19 in Airborne de luxe professional mower. The engine controls and snorkel air filter were mounted on the handlebar well away from the rotor hood.

5.63. The 18 in cut Atco rotary mower with a 145 cc four-stroke engine cost £35 in 1969.

5.65. *The Atco 11/36 E ride-on rotary mower had an 11 hp four-stroke engine.*

Five Atco ride-on rotary mowers and lawn tractors were current in the early 1980s. The 32, 36 and 42 in cut models had 7 hp Briggs & Stratton engines. Apart from a basic 32 in cut model the other Atco ride-on lawn tractors had electric starting. The cut grass was ejected from the rear or collected in a large towed grass box. Atco products in the late 1980s included the earlier lawn tractors, the Atco Clearway B30 self-propelled rotary mower and four models of the Stoic rotary mower.

BARFORD

The Barford Rotomo rotary grass cutter cost £52 10s 0d when it was introduced at the 1956 Royal Show. The 24 in cut mower, recommended for cutting rough grass in orchards and poultry runs, had a direct vee-belt drive from a 2 hp Villiers four-stroke engine to a rotor with four double-edged blades which could cut in either direction. Cutting height was varied with a single screw adjuster linked to all four 10 in diameter solid rubber-tyred wheels.

Barfords of Belton, Lincs, also made an 18 in cut version of the Rotomo with a 98 cc two stroke engine and a single screw height adjustment, which cost £30 in 1962. A 2 ft cut hand-propelled Barford Rotomo with a four-stroke Villiers engine and a rotor with four double-edged blades was made in the mid-1960s.

BARRUS

Ernest Prouty Barrus learnt his trade in Massachusetts before crossing the Atlantic to establish a Thameside warehouse in London in 1917 from where he sold American brands of workshop tools and equipment and Johnson outboard motors. The business continued in Upper Thames Street until 1938 when it moved to larger premises at Acton in London. Johnson Outboards bought the American lawn mower manufacturer Lawn-Boy and the Canadian Pioneer chainsaw manufacturer in 1955.

The Barrus link with Johnson led in 1959 to the introduction of Pioneer chainsaws to the UK market and in 1962 to the two stroke engined Lawn-Boy rotary mowers. A new grass-collecting system called

5.65A. *The Mowmaster 19 in cut mower cost £39 10s 0d in 1969, the optional grass box adding £5 to the price.*

Direct Collect introduced in the mid-1960s was used for Mowmaster mowers designed and manufactured by Barrus and the MTD Direct Collect tractor.

The Mowmaster 19 in cut rear discharge rotary mower with a 3½ hp four-stroke engine with either manual or electric starting and optional grass box was included in the 1969 Barrus price list at £42 0s 0d or £68 10s 0d. The 14 in cut Mowmaster economy model with a four-stroke engine, recoil starter and a sheet metal deck cost £29 15s 0d in 1969.

E P Barrus moved to Bicester in 1977 from where it sold Lawnflite self-propelled and ride-on rotary mowers made by MTD (The Modern Tool & Die Company), a company founded in 1956 in Cleveland, Ohio. Barrus joined forces with MTD in 1980 when Yardman and Cub Cadet tractors, Polaris ATVs, Yanmar mowers and tractors and Victa mowers were added to the Barrus product range.

BEAVER

5.66. The Beaver 80 motor scythe.

Beaver Motor scythes made by G A Holt Ltd in Kingston-on-Thames between 1956 and 1965 were advertised as the eager Beavers. They would clear dense undergrowth of brambles, bracken, gorse and thistles with their flashing blades and cut the finest lawns to perfection even when the grass was soaking wet.

The Beaver Model 80 introduced in 1952 had a 1 hp two-stroke JAP engine with a vee-belt drive to the twin-blade 18 in cut rotor. An optional 110 volt generator for the Beaver could be used with a high-pressure spraying outfit and Tarpen hedge trimmer or chainsaw.

The 18 in cut Beaver 120 with a four-stroke Clinton engine was added in 1959 and the Beaver 40 with a two-stroke JAP 80 engine introduced in 1962 was an improved version of the Beaver 80. The 18 in cut Beaver 40 was designed on the hover principle but as suitable plastic materials were not available it had an aluminium rotor hood which proved to be less effective than subsequent makes of air-cushion mower.

The 1961 Beaver motor scythe price list included the Model 40 for £32 10s 0d, the Model 80 was £39 10s 0d, the Model 120 cost £44 10s 0d and spare cutter blades were 2s 6d each. Beaver motor scythes were discontinued in 1965.

BOADICEA

In the late 1950s H R Nash of Ashtead in Surrey made the first Boadicea lightweight rotary mowers with a tiny 34 cc two-stroke engine. The handles on the cast-aluminium rotor housing could be reversed for cutting steep banks and the sides of ditches. A 14 in cut Mk II version, which cost £23 10s 0d in 1962 and £27 in 1969, was advertised as a first-class engineering product not built for price alone and designed for all members of the family to use.

BOLENS

Most of the American-built Bolens ride on rotary mowers imported by Garden Machinery of Slough in the late 1950s were used by local authorities and professional gardeners, but some small ride-on machines were suitable for the larger domestic

5.67. Sales literature explained that the Lawn Keeper would twist its way through the most challenging obstacles with ease and worm its way along irregular borders, eliminating back-tracking and hand trimming.

5.68. The mid-1970s Howard Bolens 828 riding mower had an 8 hp Tecumseh engine.

garden. The 24 in cut Bolens Suburban rotary grass cutter, which cost £90 when in was introduced in 1959, had a rear-mounted 3 hp Lawson side-valve engine, one forward and one reverse gear and handlebar steering on its small diameter front wheels.

The early 1960s pivot-steered Bolens Estate Keeper and Lawn Keeper with front-mounted rotary mower decks could be used with a front-mounted cylinder mower and for towing a small trailer or small gang mowers. The Estate Keeper had a single-cylinder 7¼ hp Wisconsin air-cooled petrol engine, six forward gears and two in reverse, a combined clutch and brake, live power take-off and shaft drives to the rear axle and the 32 or 38 in mower deck.

The 6 hp Lawn Keeper had a Briggs & Stratton power unit, two forward and reverse speed ranges and a vee-belt drive to the 28 in cut quick-attach rotary mower deck. An advertisement for the centre-pivot steered Bolens Lawn Keeper explained that, with the operator seated up front ahead of the engine noise and exhaust, the machine could be steered wherever there was grass to mow.

The Howard Rotavator Co was appointed sole UK concessionaire for Bolens garden tractors and ride on mowers in the late 1960s. The 1970s range of Howard Bolens lawn and garden tractors with mid-mounted mower decks included the 8 hp Briggs & Stratton-engined Bolens G-8 and the Bolens 828 and 829 riding mowers with rear-mounted 8 hp Tecumseh engines, three-speed transmission and a 28 in single blade mower deck.

When the Howard Rotavator Co ceased trading in 1985, Claymore Grass Machinery, a subsidiary of Reekie Engineering in Arbroath, was appointed distributor for Bolens ride-on mowers, Eurotrac lawn and garden tractors and Estate tractors. Bolens ride-on mowers and the Eurotrac range with 30, 36 and 42 in mower decks had 8, 11 and 16 hp Briggs & Stratton engines respectively.

CATCHPOLE

The Mowmor rotary mower was introduced in 1957 by Catchpole Engineering of Stanton near Bury St Edmunds, a leading manufacturer of sugar beet harvesters. There were two models, the Mowmor Major which had a 98 cc Villiers Mk VIIF two-stroke engine with a recoil starter which cost £33 and the smaller 18 in cut Mowmor Minor which had a 70 cc Villiers two-stroke and cost £26. The Mowmor had two blades, which made up to 7,000 cuts per minute, and with three cutting height positions it could deal with vegetation up to 4 ft high.

DENNIS

The mid-1950s Dennis Swift 20 in rotary mower had a horizontal crankshaft 3 hp Villiers engine and a positive drive to the large diameter rear wheels which were set inside the cutting width. The front of the mower was carried on two small diameter rubber-tyred wheels.

Publicity material noted that the engine, which had a stout carrying handle, could be removed in less than a minute by means of spring clips and used a separate power pack for various tools including a hedge trimmer, a chainsaw and a grinder/drill head. The Mk II Swift, introduced in the mid-1960s, had a cooler running engine, an improved main drive and better-quality solid rubber tyres.

Designed for local authority use, the two-speed Dennis Swallow rotary mower which appeared in the mid-1960s had power steering and 27 in long blades with renewable tips. Optional equipment included a trailed seat and a power take-off driven flexible shaft for a range of powered hand tools. About 500 Swallows and 1,000 Swifts had been made when they were discontinued in the late 1960s.

FARMFITTERS

Farmfitters of Gerrards Cross were marketing the Bushwakka motor scythe and scrub cutter and the Rapier grass cutter in the late 1950s. The Bushwakka was designed to cut really dense growth. Mounted on

5.69. The Dennis Swift engine unit and handlebars could be removed and used to operate a hedge trimmer and other power tools.

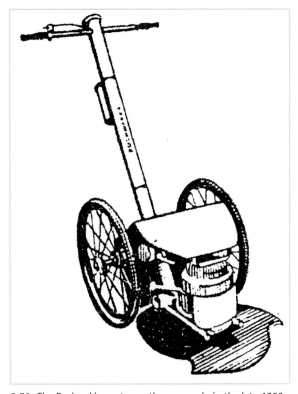

5.70. The Bushwakka motor scythe was made in the late 1950s.

5.71. The Farmfitters Kingfisher rotary was said to give every lawn a royal velvet finish.

FLYMO

A British engineer invented the Hovercraft but it was Swedish lawn mower manufacturer Karl Dahlman who designed a flying lawn mower without wheels that could float in any direction on a cushion of air. This revolutionary new machine called the Flymo was awarded a gold medal at the 1963 Brussels Inventors Fair. Great Britain was Europe's largest market for lawn mowers at the time and for that reason it was decided to manufacture the Flymo hover mower with an Aspera engine at Newton Aycliffe in County Durham. Patents were secured and a small number of hover mowers were made in 1964. Quantity production got under way in 1965 and the flourishing Flymo business was acquired by Electrolux in 1968.

large rubber-tyred wheels, an advertisement described the Bushwakka, which cost £52 10s 0d, as an extremely powerful rotary grass cutter. The 18 in Rapier rotary grass cutter with a 12½ hp Vincent engine cost £39 10s when it was introduced in 1956.

The early 1960s Farmfitters Multi-Gardener with a Clinton two-stroke or a Briggs & Stratton four-stroke engine could be used as a rotary cultivator or rotary mower. Costs in 1964 were £29 5s 0d for the engine unit complete with petrol tank and handlebars, £37 10s 0d for the rotary cultivator and £13 10s 0d for the Rapier 18 in cut rotary mower.

When it was introduced in the mid-1960s, the Farmfitters 18 in cut Kingfisher rotary mower with a two-stroke JAP engine cost £42 12s 3d. The engine had an extra large pulley for rope starting. One of the rear wheels was recessed to allow the mower to cut right up to the edge of the lawn, and the semi-pneumatic flat tread tyres were designed to leave no wheel marks on the grass.

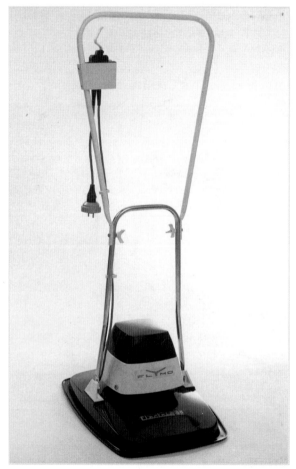

5.72. The first mains-electric Flymo hover mowers were made in 1969.

5.73. A typical late 1970s example of the petrol-engined Flymo with a two-stroke JLO power unit.

5.74. The self-propelled Flymo GXL-5 Autodrive had a 5 hp Briggs & Stratton engine and a rear grass collector.

Britain's gardens were getting smaller by the mid-1960s and their owners were becoming less energetic when it came to mowing the lawn. This led to a boom in sales of simple electric mowers and Flymo was able to cash in on this trend and the link with Electrolux provided a ready supply of electric motors. There were already eight models of the petrol-engined Flymo hover mower when the first mains-electric Flymo was introduced in 1969.

Hand-propelled, wheeled 18 in Flymo hover mowers with 3½ hp Briggs & Stratton engines appeared in the mid-1970s together with the Flymo Princess, which did not float. The Princess was a mains-electric 15 in cut cylinder mower with a plastic grass box and roller drive to give the traditional British striped lawn.

Flymo also made garden cultivators in the late 1970s and the acquisition of Norlett in 1981 added extra models of wheeled rotary mowers and garden cultivators to its product range.

The mains-electric Flymo DXE, introduced in the late 1970s, was the first hover mower with a grass collector. Ever-increasing sales of Flymo mowers led to the notorious Mower War in the early 1980s. A Flymo publicity campaign, in a sales drive against the growing popularity of small mains-electric cylinder mowers claimed that it was 'a lot less bover with a hover'. Qualcast countered this in its advertisements for the Concorde mains-electric cylinder mower with the slogan 'It's a lot less bover than a hover'. Within a couple of years it did not matter any more: the patents protecting the hover principle had expired and other companies soon added hover mowers to their product range.

In 1981 there were eighteen different Flymos. They ranged from a 10 in mower costing under £50 to a 30 in ride on Flymo with a floating mower deck and a seat on wheels for the operator which cost over £800. Flymo celebrated its twenty-first birthday in 1985 with the launch of the electric Sprintmaster hover mower. A vacuum system collected most of the trimmings in a rear grass box.

The Flymo Chevron range of wheeled electric rotary mowers with a grass collector and rear roller to give the admired striped lawn was launched in 1986. Petrol-engined Chevron mowers were added a year or so later. Flymo changed its colours from orange and brown to orange and grey and brought the hover principle up to date in 1993 with the Hovervac. The aptly named, new model had an integrated grass box in the body of the mower which was lifted out when it needed to be emptied.

5.75. Flymo introduced the Turbo Compact hover mower with an internal grass collector in the early 1990s.

HAYTER

Douglas Hayter, who started out as a building contractor in the 1930s, sold the first Hayter Motor Scythe in 1947. Having enjoyed little success in cutting rough grass with a friend's cutter bar mower he used a two-stroke motor cycle engine and tacked on various bits and pieces that came to hand to make a rotary mower. Much to his surprise it worked quite well but with a complete lack of guards it was a somewhat lethal creation with stones and debris flying in all directions.

The first production models of the 24 in Hayter Motor Scythe were made at Spellbrook near Bishops Stortford in 1947. They were simple hand-pushed machines with a 24 in wide cutting rotor vee belt driven by any suitable engine available at the time. When supplies became more readily available a four stroke Villiers horizontal crankshaft engine was used

5.76. Introduced in 1947, the Hayter Motor Scythe remained in production for twenty years.

for the Motor Scythe. There were no guards around the rotor hood and the near-side rear wheel was inset to allow grass at the side of the lawn to be cut without the wheels dropping off the edge. Sales literature claimed that coarse, tough grass difficult to cut with a scythe and impossible to cut with any other type of mower could be mastered with the greatest of ease with a Hayter Motor Scythe.

In common with most 1950s garden tractors and cultivators, various attachments including a fruit tree sprayer and a generator could be used with the Motor Scythe. The sprayer unit, which cost £21 10s 0d, had a vee belt driven pump, with a suction hose and hand lance. The belt-driven 400 watt, 110 volt DC generator, which cost £19 15s 0d, was bolted to the Motor Scythe. Recommended Tarpen electric hand tools included Hedgemaster and Grassmaster trimmers and a ½ in drill chuck. In 1951, an improved Motor Scythe with a new cutter plate and front wheels instead of the earlier skids was exhibited at the Smithfield Show.

During its early years Hayters concentrated its efforts on manufacturing agricultural mowing machines. The 5 ft cut tractor-drawn Hayter Orchard Mower appeared in 1950 to be followed by the 6 ft cut power take off driven Haytermower in 1953. The Spellbrook company turned its attention to tractor-drawn rotary mowers for highway authority use and in 1956 launched the 6 ft 6 in tractor-mounted Highway Mower. Following the success of the hand-propelled 24 in Motor Scythe, which remained in production until 1973, Hayters launched the 26 in cut self-propelled Hayter Scythe mainly for professional and highway use at the 1956 Royal Smithfield Show. The new model, which cost £108, had a 250 cc Villiers four-stroke engine, centrifugal clutch and chain drive from a three-speed gearbox to a large diameter driving roller. The vee-belt driven cutting rotor had a maximum 6 in cutting height. A trailer seat was available as an optional extra.

In 1957 Hayters entered the domestic rotary mower market with the 18 in cut Hayterette powered by a 98 cc Villiers engine with recoil starter. Many improvements have been made over the years but with well over 300,000 machines already sold the Hayterette remains in production to the present day. A variety of colour schemes have been used for the

5.77. This neatly dressed gentleman was able to cut close up to his fruit trees with this 1960 Hayter Motor Scythe. Unlike earlier models it had guards to protect his toes from the high-speed blades.

5.78. The first Hayterettes were made in 1957 and cost £36 delivered in England and Wales.

Hayterette since the original red and silver livery, including the distinctive hammer green in the 1960s, followed by red and green and then the current dark green and black first used in the early 1990s.

The Hayterette and other makes of rotary mower which returned the clippings to the ground, did not leave the traditional striped picture-book lawn. To meet the demand for a striped finish with a rotary mower the Spellbrook company introduced two versions of the Haytermower in the late 1950s. The self propelled roller-drive Haytermower designed for large areas of grass had an output of up to half an acre per hour while the 24 in hand-propelled model with a grass box under the handlebars was made for medium-sized gardens.

The 26 in cut rotary mower deck on the self propelled model with split roller drive was interchangeable with a 30 in cylinder mower attachment. The Haytermower had three gears with a top speed of 4½ mph and the rotary cutter height was adjustable from 1½ to 6 in. Optional Haytermower accessories included a wheeled or roller-mounted trailing seat, centrifugal self-priming pump, generator and a flexible drive shaft for a Tarpen hedge trimmer.

In 1976 the factory at Spellbrook was enlarged to provide extra production space for an increased range of hand propelled, self propelled and tractor-drawn rotary mowers. New models included the 18 in cut hand-propelled Haytermower with hinged cutter blades on a circular steel rotor. The cushion-tyred wheels had individual cutting height adjustment and the nearside wheels were inset to allow the mower to cut right up to the edge of the lawn.

Vacuum action was used to lift the grass and direct the clippings into a collecting bag at the rear. A deflector plate, used in place of the grass box, was provided for cutting rough grass. The 18 in Haytermower had a four-stroke Briggs & Stratton engine with a wind-up starter. Sales literature explained that the large silencer ensured quiet running, making the Haytermower ideal for use in built-up areas.

Domestic and professional models of the Hayterette S Series with a single lever to set cutting height were introduced in 1970. The professional

5.79. Late 1960s Haytermowers could be used with Tarpen flexible-drive hand tools.

Hayterette had a JLO two stroke engine while the domestic model had a 147 cc four-stroke Briggs & Stratton engine with a recoil starter. In 1967 it cost £37 which represented an increase of £1 on the price of the Hayterette when it made its debut in 1957. User safety had to be taken into account in the late 1960s when an optional guard kit, priced at £2 5s 0d, was introduced to comply with new regulations concerning the use of powered rotary mowers in agriculture.

Hayters adopted various bird names for its mowers in the early 1970s. The hand-propelled 12 in Hawk and 19 in Hawk Major rotary mowers were available with wheels or rollers. The self propelled range included the Harrier, Osprey and Condor with cutting widths from 19 in on the 3½ hp Briggs & Stratton-engined Harrier to the 30 in Condor with a 6 hp MAG engine. The hand-propelled Merlin with a

5.80. The 1970s Hayterette had red and green paintwork and a re-shaped mower deck.

5.81. The self-propelled Hayter Harrier roller mower gave a striped finish to the lawn.

3½ hp Briggs & Stratton engine and grass box was added in 1973 and, like the roller-drive Harrier, it left the lawn with the much admired light and dark green stripes.

The Bank Rider, introduced in 1974, was the first Hayter ride on rotary mower. As the name suggests it was designed to cut sloping grass surfaces including those at the side of public highways. It was one of the first rotary mowers with a hydraulic motor drive to the cutting deck and the offset operator's seat could be swivelled so that it remained level when driving across a sloping surface. About 100 Bank Riders were made, most of which were sold in Lancashire.

The 36 in cut ride on and pedestrian-controlled versions of the Hayter Frigate rotary mower replaced the Bank Rider in 1976 and the 53 in cut Eagle was

added in 1978. Designed for professional and public authority use, the Frigate was powered by a 7.3 hp Kohler four stroke petrol engine. Transmission was through a transaxle unit with three forward gears and one reverse, differential and final drive. Interchangeable front-mounted rotary and cylinder mowing decks were driven from the engine crankshaft through a hand clutch and vee belt. A drawbar was provided for trailed equipment.

The self-propelled Eagle rotary mower had three cutting rotors vee belt driven by a 16 hp Kohler engine. The transaxle transmission provided three forward gears with a top road speed of 8 mph, one reverse and a creeper gear. The 1979 Hayter price list included the pedestrian-controlled Frigate at £1,043, the ride on model at £1,253, the Eagle at £3,235 while the price of the Hayterette had risen to £150.65. The last pedestrian-controlled Frigates were made in 1980; the Eagle was discontinued in 1983, and the Hayterette was still in production in 2005.

The Hayter Hobby with a 3½ hp Briggs & Stratton engine, 16 in cutting rotor and a strong plastic grass box appeared in 1982. Designed for small and medium-sized lawns, the Hobby had a split rear roller, designed to make it easy to push and leave the banded finish so highly prized by keen gardeners. Mains-electric and Tecumseh-engined self propelled versions were added to the Hobby range in 1985.

Changes at Spellbrook saw Hayters listed on the Stock Exchange in the early 1980s. In 1984 F H Tomkins, a financial holding company, bought Hayter plc. In 1988 Tomkins then acquired the Murray Ohio Manufacturing Co which at that time was the world's largest mower manufacturer. Before the Tompkins take over the Murray range had been marketed in the UK by G D Mountfield of Maidenhead.

The change of ownership resulted in Murray rotary mowers, including a 30 in cut 10 hp rear-engined ride on machine and an 18 hp garden tractor with an underslung mower deck, being added to the Hayter

5.82. The Hayter Hunter range, launched in 1984, included hand- and self-propelled 16, 18 and 21 in cut machines.

range. Beaver Equipment, which made grass machinery for golf courses, parks and amenity areas, was bought by Tompkins in 1987, and from 1991 Beaver grass equipment was made at Spellbrook.

In 1996 Hayters launched three models of the Jubilee rotary mower with 17 to 19 in cutting widths to celebrate its fiftieth anniversary. Ten years later a full range of Hayter mowers from the hand-propelled Hayterette to the 17½ hp Heritage ride-on rotary mower was being made at Spellbrook. In 2004 the American grass machinery manufacturer Toro acquired the Hayter business.

HELI-STRAND

Heli-Strand Tools, based at Rye in Sussex, sold the Heli-Swift 20 in rotary grass mower in the early 1960s. Suitable for lawns, rough grass and scrub, the hand- and self propelled Heli-Swift rotary mowers were made with the choice of a two or four stroke engine. An unusual feature of the Heli Swift 20 in mower was that with the drive belts removed the engine could be lifted from the mower without tools and used as a portable power pack with various hand tools, including a log saw, a hedge trimmer and a hand-held rotary tiller.

LADYBIRD

Small electric rotary mowers, similar to those now sold in their thousands, were being made in the late 1940s. The Ladybird mains-electric rotary mower, made by AMI Lawnmower Co of London had a single handle, front roller and high-speed blades which minced the grass so finely that raking and sweeping was unnecessary. Advertisements at the time pointed out that the Ladybird, which cost 16 guineas (£16 16s 0d) in 1950, was as easy and economical to use as a domestic vacuum cleaner. Made mainly of aluminum alloy the Ladybird weighed a mere 16 lb and was suitable for AC or DC mains supply. It was said to be so manoeuvrable that a child could use it in complete comfort and safety.

LANDMASTER

Landmaster, based at Hucknall in Nottinghamshire, made its own range of mowers as well as importing Stoic rotary mowers in the 1960s and 1970s. The 19 in cut side-discharge Landmaster Stoic introduced in

5.83. The Landmaster Stoic rotary mower, which cost £47 10s 0d in 1963, had an unusual cutting height adjustment and an optional side-mounted grass collector.

1963, which could be used with an optional grass box or windrowing chute, had a 3 hp Briggs & Stratton engine with a wind-up starter.

The 14 in cut Landmaster Rotary 14 and the 19 in cut Stoic were the current models in the late 1960s. The 3 hp hand-propelled and 3½ hp self-propelled Landmaster 14 both with Briggs & Stratton engines had a front grass box. The rear roller left a striped finish to the lawn. The hand- and self-propelled 19 in cut side-discharge Landmaster Stoic mower had a 3½ hp Briggs & Stratton engine and a heavy-duty three-blade cutting rotor. The self-propelled Stoic had a separate clutch control for the driving wheels.

The 1970 Landmaster rotary lawn mower price list included the Saturn, Spartan and Stoic 19 with prices ranging from £34 15s 0d for the 14 in cut hand-propelled Saturn to £72 10s 0d for the 3½ hp self-propelled Stoic 19. The 3 hp hand-propelled and 3½ hp self-propelled Saturn roller mowers with a front grass box were improved versions of the Landmaster Rotary 14.

The 16 in cut hand-propelled Spartan with a 3 hp Briggs & Stratton four-stroke engine and recoil starter had folding handlebars and an optional grass bag. The Saturn and Stoic were the current models when Landmaster moved to Poole in Dorset in the early 1970s.

The Saturn, Scout, Sovereign de luxe and four models of the Mk II Stoic 19 were included in the 1978 Boscombe Engineering lawn mower catalogue.

The hand-propelled Saturn had been discontinued by then and a more powerful 4 hp Briggs & Stratton four-stroke engine was used for the self-propelled roller-drive 14 in cut model with a rear grass box. The hand-propelled 18 in cut Scout rotary mower with twin swinging blades had a 3½ hp Briggs & Stratton engine and a rear polythene grass box. The 18 in cut self-propelled Sovereign de luxe with front and rear rollers designed to give the lawn a banded effect was powered by a 4 hp Briggs & Stratton engine with a snorkel air-filter and a deadman's control lever on the handlebars.

The 19 in cut side-discharge Landmaster Stoics had three-bladed cutting rotors under a diecast aluminium deck. The most expensive self-propelled model had a 4 hp Briggs & Stratton four-stroke engine.

Following the acquisition of Landmaster in 1980

5.84. The front and rear rollers on the Landmaster Saturn gave lawns a striped finish.

5.85. The Landmaster Scout had a dished rotor with two swinging blades.

the Wolseley Webb catalogue included the Landmaster Saturn, Scout and Sovereign de luxe and three models of the 19 in cut Stoic rotary mower. Wolseley Webb introduced a new version of the 18 in cut Landmaster Sovereign de luxe self-propelled mower in 1982. The specification included a 4 hp Briggs & Stratton engine, a clutch control for the chain drive to the rear roller and a single-lever cutting height adjustment.

The Sovereign de luxe rotary mower was discontinued in 1984 but the 1985 Atco price list and the Suffolk Lawnmowers list for 1986 included 3½ hp hand-propelled and 4 hp hand- and self-propelled B 19 Stoic rotary mowers with Briggs & Stratton engines.

LAWN-BOY

E P Barrus, based at Acton in London, introduced the American Lawn-Boy rotary mowers to the UK market in 1962. The hand- and self-propelled 18 and 21 in cut Lawn-Boy mowers had a vacuum system to lift the clippings into a cloth grass catcher. The hand-

5.86. Lawn-Boy mowers were made in America.

propelled 18 in cut model cost £32 10s while the 21 in Lawn Boy Automower complete with grass bag was £52.

Mid-1970s Lawn-Boy rotary mowers included hand- and self-propelled 19 and 21 in cut side-discharge, grass collecting and mulcher models with two-stroke engines. The 5 hp four-stroke 25 in cut Lawn-Boy 25 Rider and the 36 volt battery-electric 19 in cut mower with fold-down handlebars completed the range.

E P Barrus moved to Bicester in 1977 from where it distributed the Lawn-Boy range until the early 1980s when they were marketed by NJB Mowers of Downham Market in Norfolk. A 127 cc two-stroke engine with electronic ignition was used for most mid-1980s models of hand- and self-propelled Lawn-Boy rotary mowers made for home and professional gardeners.

5.87. The Lawnflite riding mower had a top speed of 7 mph.

LAWNFLITE

E P Barrus imported a range of Lawnflite self-propelled and ride-on rotary mowers, made by MTD at Cleveland in Ohio in the early 1970s. The cheaper rear-engined ride-on mowers with a 5 or 8 hp Briggs & Stratton engine had a single forward and reverse transmission or an automatic drive system with cruise control and a top speed of 7 mph.

A direct-collection system for grass clippings, introduced by E P Barrus for Lawnflite pedestrian-controlled and ride-on mowers, used a large turbo fan which combined with the cutting blade rotor to throw grass, leaves and other debris straight into the grass collector.

The Lawnflite 900 series ride-on mowers had a twin-blade 40 in rotor while the smaller 600 Series was equipped with a single-blade cutting rotor. Both had a Briggs & Stratton engine, an automatic variable speed transmission and a direct grass collection system.

Mid-1980s Lawnflite rotary mowers included 8 to

16 hp garden tractors with 26 to 44 in mower decks and grass collectors. The 16 to 22 in cut hand- and self-propelled pedestrian-controlled models had 3½ to 5 hp Briggs & Stratton engines.

LOYD

The hand-propelled Loyd Motor Sickle made by Vivian Loyd & Co in Camberley, Surrey cost £39 15s 0d when it was introduced in the late 1940s. The 18 in cut Motor Sickle had a 98 cc Villiers two stroke engine with a vee belt drive to the 700 rpm twin-blade cutting rotor. Although the Motor Sickle was mainly used to cut rough grass and scrub, the mower disc could be replaced with a 14 in circular saw blade, which the makers claimed could cut down trees and saplings of up to 8 in diameter. A cutter bar version of the Loyd Motor Sickle was made in the early 1950s.

Publicity material stated that the Loyd

5.88. The Loyd Motor Sickle was exhibited at the 1950 Royal Smithfield Show.

Motor Sickle used no more than a gallon of petrol in an eight-hour day, that the operator could work all day long without fatigue and required no mechanical knowledge to operate the machine. It was also pointed out that blunt cutter blades could be sharpened in position or replaced in less than three minutes at a trifling cost. One satisfied user wrote that the Loyd Motor Sickle not only cut coarse grass, garden produce, brambles and undergrowth but it also cut labour, time and costs, everything in fact except your taxes and your hair!

MAYFIELD

An 18 in cut front-mounted rotary mower was among the optional attachments made by Mayfield Engineering in Croydon for the Croft Mayfield Mk 15 motor scythe. Introduced in 1955, and built for ten years, the 3 hp Mk 15 had three forward gears, an

optional reverse gear, roller chain drive to the wheels and the rotary mower was vee-belt driven. The same rotary mower attachment could also be used with the 2 hp Croft Mayfield Mk 14 motor scythe.

A 25 in cut front rotary mower attachment was an optional extra for the 4 hp Croft Mayfield Mk 20 garden tractor made from 1958 to 1965. It had a three forward and one reverse gearbox with roller chain drive to the wheels while the cutting rotor with four double-edged blades was vee-belt driven. When one cutting edge of the blades became blunt the vee-belt could be twisted to reverse the rotation of the rotor in order to bring the opposite cutting edges into use. Other attachments for the Croft Mayfield Mk 20 garden tractor included a 24 in wide cylinder mower, a 24, 36 or 48 in wide cutter bar and a hedge trimmer.

Launched in 1961, the Mk 21 pedestrian-controlled rotary mower with a three-bladed rotor disc and a 1 ft

5.89. The Allen Mayfield Merlin cost £248 in 1966. (Bill Castellan)

5.90. The Merlin rear wheels were chain driven.

9 in width cut was the only Croft Mayfield pedestrian-controlled tractor made specifically for cutting grass. It had a 4 hp four-stroke engine, two forward gears and a free-wheel mechanism to facilitate easy cornering. Optional equipment included a 2 ft cutter bar and a flexible shaft driven from the engine power take-off pulley for use with a hedge trimmer and other work heads.

The Mayfield Hoe & Mow introduced in 1961 and the later Arun Hoe & Mow (page 166) were dual-purpose machines used with a toolbar or a front-mounted 18 in cut rotary mower. The Arun Hoe & Mow was made from 1975 to 1981 by Riverside Precision & Sheetmetal at the old Croft Mayfield works at Littlehampton in Sussex. Attachments for the Hoe & Mow included a front toolbar and a vee-belt driven front-mounted 18 or 25 in cut rotary mower with a four-blade rotor.

The Allen Mayfield Merlin ride-on rotary mower was introduced in 1966 but only 109 had been made when production came to an end in 1968. Many standard Mayfield tractor components were used to build the Merlin which had a 6 hp Briggs & Stratton engine, a twin plate clutch, a three forward and one reverse gearbox, a roller chain to a differential rear axle and shoe brakes on the rear wheels. Controls included a hand throttle, clutch and brake pedals and a hand lever to adjust the four-blade aluminium rotor to its maximum 3 in cutting height.

MOUNTFIELD

G D Mountfield was established in Maidenhead in 1962 to manufacture lawn mowers and garden cultivators. The M1 garden cultivator with an optional rotary mower attachment together with the M2 and M3 rotary mowers were the first Mountfield products. The rear-discharge M2 on four large rubber-tyred wheels and fold-down handlebars was designed to cut rough grass while the 18 in cut M3 with a 3 hp Australian-built Kirby engine, a two-piece rear aluminium roller and rear grass box was for domestic lawns.

The 1975 Mountfield price list included hand- and self-propelled versions of the 3½ hp M3 and five models of the 18 in cut M4. There were also the 21 in cut M5 and M6 with Aspera engines and the M35, M3 and M5 mains-electric rotary mowers. Large rubber-tyred wheels set within the cutting width of the die-cast alloy mower deck and individual cutting height adjustment for each wheel were features of mid-1970s Mountfield rotary mowers. A 2½ hp two-stroke or 3 hp four-stroke engine was used for the M4 Minor; the 3 hp M4 de luxe, the 3½ hp M4 Major and M4 Power Drive had four-stroke engines. Also made in the mid-1970s were mains-electric versions of the 18 in cut M3 and M4 rotary mowers and the 14 in cut Mountfield M35 with a 16 m or 32 m cable.

The 4 hp four-stroke engined 21 in cut self-

5.91. An early 1970s Mountfield rotary mower with an Aspera engine.

propelled Mountfield M5 had a two-speed transmission and a vacuum system to lift clippings into the grass box. The higher gear was for fine grass and users were advised to use low gear when cutting rougher areas. The two- or four-stroke engined 21 in cut self-propelled M6 completed the 1975 range of Mountfield petrol-engined rotary mowers.

In the early 1980s the Empress and Emperor roller-drive and wheeled Emblem rotary mowers with Briggs & Stratton or Tecumseh engines were added to the Mountfield range. The 15 in cut hand-propelled Emblem was designed for the smaller garden and there was a choice of a 16 in cut hand-propelled Empress or an 18 in cut self-propelled machine with optional electric starting. The 4 hp Emperor with a 21 in cutting deck completed the range of Mountfield roller-drive mowers.

Sales literature explained that the electric-start models had a sealed-for-life battery with an alternator to keep it in peak condition but an auxiliary pull-start was provided – just in case.

G D Mountfield became part of the Ransomes Consumer Division when it was acquired by Ransomes Sims & Jefferies in 1985. The Mountfield rotary mower catalogue for 1986 included hand- and self-propelled versions of the Empress, Emperor, Emblem and Mirage with Briggs & Stratton, Tecumseh and Suzuki petrol engines. The 15 in cut hand-propelled Emblem was also made with a 240 volt electric motor. The choice of engine for the 18 in cut hand- and self-propelled Mirage wheeled rotary mowers was Briggs & Stratton Quantum; overhead valve Suzuki; Tecumseh four-stroke, or a two-stroke Suzuki.

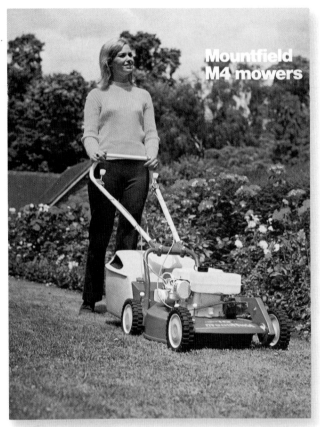

5.92. The front wheels of the 21 in self-propelled Mountfield M4 rotary were chain-driven through a two-speed gearbox.

5.93. The Mountfield M25/5 could be used with a rear grass catcher or a side-discharge chute.

The 21 in cut Monarch appeared in 1988 while the 14 in cut mains-electric Princess roller mower with a safety brake, which stopped the blade in three seconds, was added in 1989. Mountfield mowers in the early 1990s included Mirage, Monarch, Princess, Emperor and Empress models. In 1993 the Laser wheeled mowers were launched with the option of a mains-electric or Briggs & Stratton power unit.

Mountfield also imported Wheel Horse ride-on rotary mowers in the early 1970s. The range included the R26 ride-on rotary with a floating mower deck, a 5 hp Lausen four-stroke engine and a two forward and two reverse gearbox. The Mountfield Wheel Horse Commando 8 and Lawn Ranger garden tractors had mid-mounted mower decks and could be used with a variety of trailed implements. The Lawn Ranger had three forward speeds and one reverse and a single pedal controlled the clutch and brakes.

5.94. The Ransomes Mountfield Empress roller mower.

In the mid-1980s Mountfield also distributed Murray ride-on and pedestrian-controlled mowers but this arrangement ended in 1988 when Tompkins, the owner of Hayters of Spellbrook, bought the Murray Corporation.

In the early 1980s Mountfield also marketed a range of American-built lawn and garden tractors and ride-on mowers which were sold under its own name. They included the 25 in cut 5 hp Mountfield M25/5 ride-on mower and the 7 hp M25/7 with Briggs & Stratton engines, a three-speed gearbox and side-mounted grass catcher. An 8 hp Briggs & Stratton engine powered the five-speed Mountfield M30/8 ride-on mower.

The 8 and 11 hp lawn tractors, also with Briggs & Stratton engines, had electric starting, a three-speed transmission and a 30 in cut mid-mounted mower deck. Optional equipment for the Mountfield garden tractor with a twin-cylinder 16 hp engine, four-speed gearbox and transaxle included a mid-mounted 36 in cut side-discharge mower deck, roller, lawn sweeper and a trailer.

The 1988 Mountfield catalogue also included a range of American-built Simplicity ride-on lawn and garden tractors. The 12½ and 16 hp garden tractors with Briggs & Stratton engines and hydrostatic transmissions had 42 in cut mid-mounted side-discharge rotary mowers. The Simplicity lawn tractor with a 36 in cut mid-mounted mower had a 12 hp Briggs & Stratton engine and a five forward and one reverse gearbox.

NORLETT

In the early 1960s Fenter Ltd of Birmingham imported the Norwegian-built Norlett rotary mowers. The smallest 19 in cut hand-propelled Norlett V4 with a 2½ hp Aspera power unit, which cost £25 10s 0d in 1962, was advertised as the lowest priced four-stroke rotary mower on the UK market. The 2½ hp 19 in Norlett de luxe was £35 while a 22 in cut self propelled model was £65. In the mid-1970s Norlett made 16 and 19 in cut rotary mowers with 3½ hp Briggs & Stratton engines for the cheaper end of the market. Sales literature described them as tough and dependable machines with elegant handles plated for protection which could be folded down for easy storage.

Norlett sold Norlett Ariens rotary mowers from Thame, Oxfordshire, its base in the 1970s. The range included the 5 and 7 hp 26 in cut Fairway, the 30 in cut 8 hp Emperor and the 10 hp Emperor de luxe with a 36 in mower deck. Flymo of Newton Aycliffe in County Durham acquired Norlett in 1981.

QUALCAST

5.95. The 2 hp JAP engine on the 18 in cut Qualcast Rotacut used less than a pint of petrol in an hour.

The 18 in Rotacut, introduced in 1957, was the first Qualcast rotary mower. Advertised as the modern way to cut grass, the Rotacut had a two stroke JAP engine with a recoil starter, safety clutch and four reversible blades. An improved Mk IV Rotacut with a four stroke Clinton engine cost £29 8s 0d when it appeared in 1962. It had offset wheels and the cutting height was adjusted with a single lever. The Mk V Rotacut with a four stroke Aspera engine announced in 1963 was still being made in the late 1960s when it cost £33 19s 6d.

The 12 in cut Rota Mini introduced in 1970 was the first Qualcast mains-electric rotary mower. It had a rear roller for mowing over the edge of lawns, a thermal cut out to protect the motor and a built in TV interference suppresser. The Rota Mini, complete with 50 ft of cable, cost £11 9s 6d.

The 14 in cut mains-electric Rota Mo 360, added to the Qualcast range in 1971, cost £25 with a grass box and 72 ft of cable. The Rota Mo was restyled in 1974 and, with a rear-mounted grass box, was re launched as the Qualcast Jetstream. The cheapest 15 in cut Jetstream with a 1,000 watt mains-electric motor and 50 ft of cable or a two stroke petrol engine cost £50.95.

The 18 in cut Jetstream with the choice of a 2½ or 3 hp Aspera engine was made for the larger garden and the 3½ hp de luxe Jetstream with fold-down handlebars completed the range. A test report on the electric Jetstream concluded that it was impossible for users to get their feet caught under the rotor cover and it was also noted that a red light on the motor which did not go out until the blades were stationary was a useful safety device.

Birmid Qualcast marketed the Atco, Qualcast and Suffolk brands in the early 1970s. In 1974 the Qualcast Jetstream rotary mower was renamed the Suffolk Meteor. The Jetstream name came back into use in 1979 when Birmid Qualcast announced three new 18 in cut petrol-engined Qualcast Suffolk Jetstream rotary mowers to replace the Suffolk Meteor and the Atco range of side-discharge rotary mowers.

The 18 in Qualcast Suffolk Jetstream and the new 16 and 19 in cut Qualcast Airmo hover mowers were introduced for the larger domestic garden while the mains-electric Qualcast Rota Safe and Mow N Trim rotaries introduced at the same time heralded a new era of mowers for the care of small lawns. The 12 in cut Rota Safe lived up to its name with a child-proof safety switch in the handle. Flexible plastic blades reduced the risk of accidentally cutting through the cable. Plastic cutters and nylon line supplied with the dual purpose Mow N Trim could be clipped to the rotor either for mowing a lawn or for trimming around trees and shrubs.

In the mid-1980s the Qualcast Suffolk Lawnmowers range of rotary mowers included the Rota Safe, the petrol-engined Jetstream, the Mow N Trim and the Atco B45 mulcher mower, together with the mains-electric Qualcast Airmo and the petrol-engined Atco Airborne B16 hover mower. The 1985 Qualcast price list also included 3 and 4 hp Stoic wheeled rotary mowers with four-stroke petrol engines.

Marketing arrangements for Atco, Qualcast, Suffolk Webb and Wolseley products changed at frequent intervals from the mid-1970s. In 1985, for example, Atco and Webb marketed their own machines and Birmid Qualcast looked after the Qualcast and Suffolk brands. Two years later Suffolk Lawnmowers were selling the Atco, Webb and Stoic brands, while the Panther, Concorde and Suffolk Punch were included in the Birmid Qualcast catalogue. Manufacture of Atco, Webb and Stoic mowers was based at Stowmarket in the late 1980s when Qualcast Garden Products in Derby was responsible for Qualcast and Suffolk grass care equipment.

RANSOMES

The Ransomes Cyclone 18 in rotary mower with a 2 hp JAP two stroke engine made in the mid-1950s was described in sales literature as the mower with a hummock disc. It had a convex guard on the underside of the two-blade rotors which, it was claimed, would ride over uneven ground without damaging the blades and prevent long grass winding around the rotor shaft and stalling the engine.

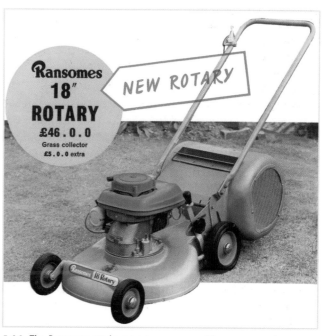

5.96. The Ransomes 18 in cut rotary mower with a rear grass box was launched in 1965.

5.97. Ransomes Mk II Multimowers could be used with a rotary mower deck or a 30 in cut cylinder mower.

The 18 in cut Ransomes Typhoon with a two or four stroke engine, costing £21 15s 0d and £26 15s 0d respectively, replaced the Cyclone in 1961. A magazine advertisement at the time suggested these were very low prices indeed, but in spite of this price advantage the Typhoon was withdrawn within a few months to be replaced by the 18 in Typhoon Major with a Villiers 150 cc four stroke engine. It was not a simple task to alter the cutting height of rotary mowers in the early days and the Typhoon was no different from most of its competitors in this regard. Spanners and sometimes a hammer were needed to alter the position of the wheels on the rotor housing.

The Typhoon and earlier Cyclone rotary mowers returned the clippings to the ground but the new Ransomes 18 in Rotary announced in 1965 had a large metal grass collector and a single lever provided instant cutting height adjustment. The Ransomes 18 in Rotary was only made for two years but the Typhoon Major remained in production until 1969.

Ransomes entered the professional rotary mower market in 1960 with the dual-purpose Multimower for cutting large areas of grass. It had a JAP 288 cc four stroke engine, two-speed gearbox and interchangeable 27 in cut rotary and 30 in three-blade cylinder mower attachments. In 1961 an improved Mk II Multimower with the same JAP engine followed, and in 1969 a Mk III machine with a MAG engine was introduced. The Multimower

2000, still with the interchangeable cylinder and rotary mowers, appeared in 1972. It was made until the-mid 1980s when Ransomes introduced a new range of professional rotary mowers and the Multimower became a 30 in cut cylinder mower.

Ransomes withdrew from the domestic market in the mid-1970s to concentrate on professional grass care equipment. However, the acquisition of G D Mountfield at Maidenhead and Westwood Engineering at Plymouth in 1985 gave the Ipswich company a renewed interest in domestic machines. A new 18 in cut heavy-duty, hand-propelled rotary, based on a mower originally designed by Mountfield, was launched by Ransomes in 1989. It had a 5 hp MAG Kubota engine mounted on a cast-alloy deck with an optional rear-mounted grass collector.

Having sold the agricultural side of the business to Electrolux in 1987 Ransomes widened its interests in professional turf care equipment with the acquisition of a number of specialist American manufacturers, including Cushman, Jacobsen and Ryan. In turn, Ransomes, Sims & Jefferies was acquired by the American Textron Corporation in 1998, a full 166 years after the first Budding lawn mower was made in Ipswich.

ROTOSCYTHE

The world's first rotary mower, the Rotoscythe, was made by Power Specialities at Slough in 1933. The original machine was the result of an attempt by David Cockburn to make a rotary hedge trimmer which he combined with a domestic vacuum cleaner to cut his hedges and collect the trimmings. It was not a success but as he dragged his invention across a lawn with the motor still running he noticed that as well as cutting a strip of grass it also picked up the lighter grass clippings. A prototype was made and after patents were taken out to protect its vacuum collection system the Rotoscythe, with a rear-attached grass box, went into production.

Power Specialities moved to larger premises at Maidenhead in 1936. Within three years the Rotoscythe range included the petrol-engined 14, 16, 18 and 20 in cut hand-propelled mowers, a 16 in cut self-propelled model and a 14 in cut mains-electric Rotoscythe. The 18 and 20 in cut machines had large diameter spoked-wheels with pneumatic tyres.

Rotoscythes were made until the outbreak of war. Shortly after production resumed in the late 1940s Power Specialities introduced the new Rotoscythe Model 16S. An advertisement explained that the new mower with 'a very wide range of usefulness was of compact form and handy in operation'. There was a direct drive from the 1½ hp Power Specialities two-stroke engine to the cutting rotor and to a two-section rear roller. A clutch was provided to disengage the drive to the rear roller when it was necessary to push the Rotoscythe into awkward corners.

5.98. An early Rotoscythe.

The 16 in rotor had three circular cutting blades with impeller fins to collect the clippings and blow them into a rear-mounted grass box. About 25% of each circular blade was exposed to the grass and when that section of the cutting edge was blunt the blades were rotated to bring new sharp edges into work. The Rotoscythe 16S cost £53 9s 0d in 1949 and a set of three moss removal pegs which cost 5s 10d could be fitted to the rotor in place of the cutting blades.

A more robust 18 in cut County Rotoscythe designed for the market gardener and smallholder cost £28 when it was launched in 1950. The County did not have a grass box and returned the clippings to the ground, a fashionable trend in mower design that proved to be short-lived. Within a few years a grass box was standard equipment on most rotary mowers including the 18 in County, 16 in Windsor and 14 in cut Eton Rotoscythes.

The Rotoscythe patents were extended beyond the normal term because of the war but when they eventually expired in 1952 it didn't take long for other companies to introduced similar designs of rotary mower. J E Shay of Basingstoke, which acquired Power Specialities in 1952, carried on making the Rotoscythe and,

5.99. The Rotoscythe 16S could also be used to remove moss.

5.100. Power Specialities Ltd introduced the County Rotoscythe in 1950.

SUFFOLK

The 18 in cut Polo, the first rotary mower made by Suffolk Iron Foundry, cost 25 guineas in the mid-1950s. The 18 in cut Centaur rotary mower with a Clinton Gem four-stroke engine and reversible blades, which superseded the Polo in 1959, remained in production until 1963 when the Suffolk Galaxy was launched. The Galaxy had a two stroke Aspera engine, offset wheels, an 18 in rotor with reversible double-edged blades and a grass collector.

The 18 in Suffolk Meteor, announced in 1974, cost £58.95 with a 3 hp Aspera four stroke engine while the 3½ hp Briggs & Stratton-engined de luxe Meteor was £65.95. The Suffolk Meteor, which was voted the best buy in its class by the French Consumers' Association, had the same colour scheme and an identical cutting performance as the Qualcast Jetstream.

The Suffolk brand name was not used for rotary mowers from the late 1970s until the early 1990s when five models of Suffolk Punch rotary mower

following the introduction of the Shay Rotogardener in 1954, the same Shay two stroke engine was used for both a rotary cultivator and the Rotoscythe. A 15 in cut Rotoscythe was added in 1957 and after a while J E Shay changed the name of most models of Rotoscythe to the Rotor Mower. Wolseley Webb took up production of the Rotoscythe when it acquired J E Shay in the mid-1960s.

SCIMITAR

Pressure Jet Markers, based at Stamford Brook in London, made three models of the Scimitar rotary mower in the early 1960s. The 8 in cut Scimitar 8 with a 34 cc two-stroke engine cost £25. The later Scimitar 18 and Scimitar Major were 18 in cut mowers with 79 cc two-stroke engines.

5.101. The Qualcast Suffolk Meteor rotary mower.

were included in Atco-Qualcast price lists. The Suffolk P16, P18 and P19 rotary mowers had Briggs & Stratton engines with recoil starters, 16, 18 and 19 in cutting widths, folding handlebars and grass collectors.

TEAGLE

The Teagle Jetmow rotary lawn mower with the well-tried Teagle 49 cc two-stroke engine and an automatic rewind starter was introduced in 1956. The domestic 16 in cut model with a rear grass box and cutting height adjustable from ¼ to 2¾ in cost £25 plus £5 12s 6d purchase tax. The 18 in cut Teagle Jetmow rotary mower, which discharged the grass sideways had a 1¼ to 5½ in cutting height and as there was no purchase tax it only cost £22. Both machines had a one-piece cutter blade secured by two bolts to the rotor shaft which was lubricated by an oil-saturated felt seal at the top of the rotor shaft.

The 16 in cut Jetmow had a grass box, two rubber-tyred front wheels and wooden rollers at the rear. The side-discharge mower had a separate cutting height adjustment for the front and rear wheels. Sales literature explained that if the engine required an overhaul it could be posted back to the makers and they would return it the next day for a maximum charge of £4.

VICTA

In 1962 Victa (UK) Ltd, based at Watford, introduced the Australian-built hand-propelled 18 in cut Victa rotary mower. It had a 125 cc four stroke engine with a wind-up impulse starter, two swinging cutter blades under an aluminium deck, eleven cutting height settings and folding handlebars. Two paddle blades on the cutting rotor discharged the clippings sideways, either on to the ground or into a side-mounted grass catcher. The mower cost £45 and the optional side-mounted grass catcher was an extra £5 5s 0d. The features of the Victa mower

5.102. The first Teagle Jetmow rotary mowers were classed as commercial machines and were exempt of purchase tax.

5.103. The Australian-built Victa VC 160 was introduced to the UK market in 1974.

were not unique but very few of the other makes of rotary mowers on the market in the early 1960s had all of them.

The Victa Commander, Corvette and Consul rotary mowers were listed in the 1964 catalogue. The self-propelled Commander with a four-stroke engine and rear grass box which cost £65 was the most expensive model. The two- and four-stroke Corvette also had a rear grass box. The Consul, also available with a two- or four-stroke engine, was a side-discharge mower with an optional side-mounted grass catcher.

Victa (UK) had moved to Ashtead in Surrey by the mid-1970s when its catalogue included five rotary mowers with rear grass catchers and three side-discharge machines, all with two stroke Victa engines. Features of Victa rotary mowers included a snorkel air filter on the folding handlebars to ensure a supply of clean air to the engine and a magic eye indicator to warn the user that the grass box was full.

Mid-1970s Victa mowers included the 125 cc 18 in cut Chevron which cost £73 and the 160 cc Super 24 professional rotary which at £182 was the most expensive model. Two-stroke Victa engines were also used on the side-discharge 18 in cut Chevron and Professional 160 and the 20 in cut Super 24.

Victa mowers with a rear grass box included the VC 125S with a 125 cc two-stroke engine, the self-propelled Victa Autodrive with a 160 cc power unit and the hand-propelled Sports and Mustang mowers. The 18 in cut hand-propelled Victa Impala had a 148 cc four-stroke Briggs & Stratton engine.

After moving to Basingstoke in the mid-1980s Victa (UK) marketed an expanded range of hand- and self-propelled Victa professional and domestic rotary mowers. The professional models included 18 to 28 in cut rear- and side-discharge mowers with Victa or JLO two-stroke engines. Apart from the Silver Streak 4 and Charger 4 with 3½ hp four-stroke Briggs and Stratton engines the other 18 in cut domestic models had two-stroke Victa engines.

E P Barrus at Bicester was appointed UK concessionaire for Victa mowers in the early 1990s when its catalogue included a full range of hand- and self-propelled professional and domestic rotary mowers. A new generation of two- and four-stroke Victa engines powered the Pacer, Commando and Mustang rear-discharge domestic models with plastic grass collectors.

WESTWOOD

Westwood Engineering at Plympton near Plymouth, which started out in a small way in 1969 manufacturing pedestrian-controlled rotary mowers, introduced the Westwood Lawnbug ride-on mower in the early 1970s. The 24 in rotary mower was powered by a 5 hp Aspera engine and could work at speeds of up to 8 mph. The wider rear track could be adjusted for mowing right up to the edge of a lawn.

The early 1980s Westwood 18/4 hand-propelled 18

5.104. The 18 in cut Westwood 18/4 had a four-stroke Briggs & Stratton engine.

in cut rotary mower had a 3½ hp Briggs & Stratton engine, a tempered spring steel blade, independent adjustment on all four 7 in diameter wheels and tubular folding handlebars. The 18/4 was recommended for gardens with uneven slopes, tricky corners, neglected areas, trees, walls and fences.

In the mid-1980s Westwood concentrated on the production of garden tractors with mid-mounted mower decks. The range included the 6 or 10 hp S series and the 11 or 16 hp T series garden tractors with petrol engines and the D series model with a 12 hp diesel engine. Cutting widths varied from 30 to 42 in and the seven D series models could also be used with a rear grass collector and a mounted sprayer with hand lance. Westwood pedestrian-controlled rotary mowers in the late 1980s included self-propelled 18 and 21 in cut models with the choice of a Briggs & Stratton or Tecumseh engine with electric starting. In 1989 Westwood Tractors was acquired by Ransomes, Sims and Jefferies when, along with G D Mountfield of Maidenhead, the two companies were merged to form Ransomes Consumer Ltd.

Briggs & Stratton petrol engines in the 10 to 18 hp bracket were used in the mid-1990s for the T, S and 2000 series Westwood lawn tractors. The new Lawnrider had a 10 hp engine with electric starting at the rear, a 30 in cutter deck and a vacuum grass collector. The more recently launched Westwood 2000 series garden tractors had a hydrostatic transmission. The improved Westwood S and T series garden tractors were more powerful than the earlier models.

WOLSELEY WEBB

When H C Webb joined the Wolseley Hughes Group in 1963 the Webb side of the organisation concentrated on lawn mower production while Wolseley handled the Merry Tiller side of the business. Introduced in the mid-1960s, the Wolseley Clearway professional rotary and cylinder mowers were a direct competitor to the Ransomes Multimower. The 30 in cut Clearway rotary mower had an 8 hp Kohler engine with a clutch to control the forward and reverse primary drive, a two-speed gearbox and roller chain drive to the wheels. A separate clutch-controlled vee-belt drive was used for the cutting rotor. The Wolseley Clearway reel mower used the same engine and transmission system to drive the 30 in cut three-blade mower cylinder.

The mid-1970s Wolseley Clearway HS 30 in rotary mower with a 10 hp Briggs & Stratton engine and hydrostatic drive had a top forward speed of 5 mph and 2 mph in reverse. A 9 hp Briggs & Stratton engine was used to drive the Clearway HS reel mower with a 30 in four-blade cylinder. Other mid-1970s self-propelled Wolseley rotary mowers included the Wolseley 500 together with standard and professional versions of 23 in and 27 in cut machines for rough grass.

The 19 in cut Wolseley 500 side-discharge rotary mower had a 4 hp Aspera with deadman's handle clutch control and folding handlebars.

5.105. A maximum slope of 30 degrees was advised when using the self-propelled Wolseley 500 rotary mower.

5.106. The Wolseley 27 in cut Professional rotary mower used about four pints of petrol an hour when cutting rough grass.

Briggs & Stratton 5 and 7 hp engines were used for the self-propelled Wolseley 23 and 27 in cut rotary mowers for rough grass, while the professional models had 5 and 7 hp Tecumseh engines. A two forward and one reverse gearbox was used on the 27 in cut standard and professional mowers and the 23 in cut machines had a top speed of 3 mph in forward and reverse.

In the late 1970s Wolseley Webb made a range of ride-on mowers and garden tractors. They included the 34 in cut 348 ride-on mower with an 8 hp Briggs & Stratton engine which had a four forward and one reverse gearbox on the rear axle. The gold-and-

cream coloured 30 in Wolseley 308 ride-on was similar but it had automatic drive with a choice of six speeds between 1 and 7 mph.

The Landmaster Stoic 19, the 14 in cut Saturn and the 18 in cut Scout and Sovereign rotary mowers with 3½ or 5 hp engines were added to the Wolseley Webb price list when it acquired Landmaster in 1980. The 30 in cut Clearway with a 10 hp Briggs & Stratton engine and hydrostatic transmission completed the 1980 range of Wolseley rotary mowers. The Landmaster and Stoic machines were discontinued in 1984 and the last hydrostatic drive Wolseley Clearway professional mowers were made in 1988.

Motor Scythes

Horse- and tractor-drawn reciprocating knife mowers had been made for the best part of a century when the first hand- and self propelled mowers, usually with a 3 ft cutter bar, arrived on the gardening scene in the early 1930s. A few horticultural cutter bar mowers are still made but in common with the agricultural industry rotary mowers have replaced them. Cutter bar attachments were available for a number of two wheel pedestrian-controlled garden tractors in the 1940s and 1950s, including the Howard Bantam, Gravely, Monrotiller and Trusty Earthquake.

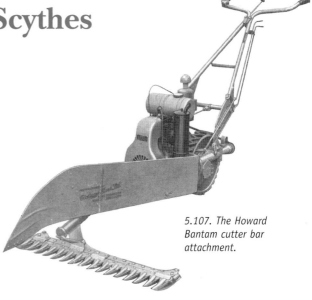

5.107. The Howard Bantam cutter bar attachment.

5.108. The cutter bar attachment for the Series III Monrotiller.

5.108A. A cutter bar mower modified for the 1930s Rowtrac.

About ten years later a 3 ft 6 in mid-mounted cutter bar with its own engine was made for the tricycle-wheeled Trojan Monotractor and a front-mounted cutter bar was included in the list of attachments for the Italian Bertolini and the Danish Texas garden cultivators. In the early 1980s cutter bars were still listed as optional equipment for Gravely, Howard Dragon, Iseki and other garden tractors.

ALLEN

John Allen & Sons of Oxford made the 3 ft cut self-propelled Allen Motor Scythe for nearly forty years. The first Allen Motor Scythe, introduced in 1933, had a 1½ hp two stroke Villiers 11C engine, friction clutch and spoked iron wheels. Improvements in 1935 included a redesigned gear housing with a dog clutch drive arrangement to the pneumatic-tyred wheels. The Mk II Allen Motor Scythe, introduced in 1937, had a modified knife drive with a spring steel rocker arm to reciprocate the knife and hold it firmly against the cutter bar fingers.

Worm-and-wheel gears combined with a pair of reduction gears were used to drive the large-diameter pneumatic-tyred or optional iron wheels with ratchets in the wheel hubs. The knife rocker arm was located in a socket on the knife back and driven from the worm shaft by an enclosed crank mechanism. An overload safety clutch protected the knife drive mechanism.

5.109. An early 1950s advertisement for the Allen Motor Scythe.

The Allen Model T Motor Scythe, introduced in 1943, had a rope-started two-stroke Villiers 25C engine with a cast-iron cylinder and piston. The engine was rarely any trouble but it often refused to start if there was more than a trace of fuel mixture in the crankcase. Draining the fuel from a drain plug under the crankcase usually solved the problem. The same Villiers 25C engine was also used for a while for the Model TS which replaced the Model T in 1952. Like the earlier Villiers engines the 25C was cooled by a built-in blower. A governor was used to control engine speed and an oil bath air cleaner was available for work in very dusty conditions.

Alternative four-stroke engines for the TS included the Villiers 3 hp Mk 15 and 4 hp Mk 25 petrol engines and the Villiers Mk 25 four-stroke petrol/tvo engine. A 4 hp four-stroke Kohler petrol engine was used on later Model TS Motor Scythes.

To make space for the optional, and later standard, forestry guards the 2 ft 3 in overall wheel width of the TS

was six inches wider than the Model T Motor Scythe. An alternative 3 ft long axle was recommended for added stability when working on slopes and the longer axle could also be used to adjust the track width when working in rowcrops with the optional front toolbar.

The Allen Universal Motor Scythe Model F, introduced in late 1957, had the same basic specification as the TS but added a new plug-in implement attachment system, a front power take-off shaft from the main drive, adjustable height handlebars, increased ground clearance and an optional three-speed gearbox. Two steel arms on the implement were plugged into the front of the scythe and the drive shaft for the cutter bar and other powered equipment was connected to the power drive shaft in a similar way.

The plug-in system provided a quick and easy method of fitting a wide selection of front-mounted implements made specifically for the Model F but attachments for earlier motor scythes were not suitable. The Model F was discontinued in the mid-1960s and the last Model TS Allen Motor Scythes were made in the early 1970s.

The 3 ft centrally mounted cutter bar was standard but 2 and 4 ft wide cutter bars were available and the 3 ft bar could be offset to either side on the machine. The Allen was much more than a motor scythe for cutting grass and scrub. Front-mounted attachments included a plough, toolbar, hay sweep, rotary brush, snowplough, yard scraper and a cylinder mower. The toolbar was bolted to the cutter bar bracket and could be used to hoe, cultivate and drill seeds.

5.111. The Allen TS Motor Scythe replaced the model T in 1952.

5.110. The Allen scythe with an offset cutter bar.

5.112. Attachments for the Allen Universal 'Plug-in' did not fit other models of the Allen Motor Scythe.

Ancillary equipment for hedge trimming, chain-sawing and sheep shearing could be used with either an engine-driven 400 watt generator or a flexible shaft vee belt driven from the crankshaft pulley. Orchard spraying was another job for the Allen Motor Scythe and a trailed seat with a drawbar was used to tow gang mowers, a hay rake and a trailer.

The Allen Junior Motor Scythe introduced in 1947 and made for two years had a JAP four-stroke air-cooled engine and a 3 ft cutter bar. Spare parts for the Junior were available from John Allen & Sons until 1961.

ATCO

The Atcoscythe with a Villiers engine to drive its autocycle-type wheel and 3 ft cutter bar was introduced by Charles H Pugh in the early 1930s. An improved Atcoscythe with a 147 cc two-stroke Atco Villiers kickstart engine, a 3 ft or optional 2 ft cutter bar and a single pneumatic-tyred wheel appeared in the late 1930s. The transmission consisted of a hand lever-operated dog clutch, totally enclosed worm-and-wheel gearing and a roller chain final reduction drive to the wheel. The cutter bar was driven from the engine crankshaft through a friction clutch, reduction gears and a vee-belt to the pitman crank.

The makers claimed that the Atcoscythe, with a work rate of between two and four acres per day, had the capacity to cut up to twelve times more grass than a

5.113. The Allen Junior Motor scythe was only made for two years.

268

The pneumatic-tyred wheels had free-wheel ratchets in the hubs and an independent front power take-off with a double disc clutch could be used to drive an optional single or twin vee-belt driven disc rotary mower, a front-mounted saw bench and a 110 volt Tarpen generator and hedge trimmer.

BARFORD

Barford Agricultural at Belton near Grantham made the Barford Councillor power scythe and a front cutter-bar attachment for the Barford Atom garden tractor. The Councillor power scythe with a 3 ft wide cutter bar, introduced in 1954, was recommended for clearing overgrown grass and weeds in gardens, orchards and paddocks. Optional attachments included a cylinder mower, a flexible drive shaft for a hedge trimmer, a snowplough and a sprayer.

The Barford Power Scythe with a 1.9 hp four-stroke engine was listed in 1958.

The cutter bar for the Barford Atom, described in a 1950 catalogue as an invaluable accessory for private gardens, nurseries and estates, cost £17 10s 0d. The 3 ft cutter bar, with optional 18 and 24 in cutting widths, was mounted centrally on the front of the tractor and an independent clutch engaged the vee belt drive from the engine crankshaft to the knife drive crank.

CLIFFORD

Clifford Aero & Auto made front- and side-mounted cutter bars for its 1950s range of rotary cultivators. A triple vee belt drive from the gearbox was used to operate the front mounted 3 ft or 3 ft 6 in cutter bar for the Model A Mk I cultivator. The knife on the 2 ft 6 in side-mounted cutter bar for Model A Mk I, Model A Mk III and Model B rotary cultivators was driven from the rotor shaft by a crank and pitman arrangement. A safety clutch protected the knife from hidden obstructions and a lever was provided to raise the cutter bar to a vertical position for transport and storage purposes.

5.114. The mid-1950s two-stroke Atcoscythe used about 1½ pints of fuel in an hour.

man with a hand scythe. Although the lack of guards over the chain drive would be totally unacceptable in today's health and safety climate, it was explained that this prevented the drive from becoming choked by trapped grass and other vegetation which would inevitably be trapped if guards were used.

Complete with a spare knife, the Atcoscythe cost £46 10s 0d in 1939, or £11 12s 6d down and twelve monthly payments of £3 2s 0d on hire purchase. Ten years later the single-wheel Atcoscythe complete with a spare blade cost £75.

A rope-started Atco-Villiers 147 cc two-stroke engine was used for the two-wheel Atcoscythe made for about ten years from the mid-1950s. It had separate clutches for the vee-belt drives from the engine to engage the drive to the wheels and the 3 ft front-mounted cutter bar. Roller chains were used to transmit power to the wheels and the cutter bar crank mechanism.

The Clifford Super Scythe, introduced in 1956, had an 80 cc or optional 147 cc two-stroke engine, two gears with top speeds of 1¼ and 2½ mph, two power take-off shafts and the choice of a 24, 30 or 36 in wide cutter bar.

CYCLO

The Cyclo Mk III Motor Scythe introduced by Cylinder Components of Birmingham in 1956 had a Villiers 2G 70 cc two-stroke engine with a Villiers Junior carburettor and flywheel magneto, tubular handles and steel wheels with solid rubber tyres. Power was transmitted by a centrifugal clutch to a vee-belt, which provided overload protection for the engine, while a tensioner pulley served as a simple clutch. The Mk III H was hand-propelled and the 22 in or optional 18 in wide cutter bar was driven by the vee-belt through a worm drive and an eccentric shaft.

The self-propelled Mk III P Cyclo motor scythe had a clutch-control lever on the handlebars and the wheels were driven by a roller chain from the cutter bar worm gearbox. The drive to the wheels could be disconnected when cutting in confined spaces and both models had a power take-off pulley for a hedge trimmer attachment. The self-propelled Cyclo Motor Scythe Mk III P with a 22 in cutter bar cost £39 15s 0d in 1956 while the Mk III H was £48.

5.115. The Clifford cutter bar attachment was suitable for the Models AI, AIII and B Clifford cultivators.

5.116. The Cyclo Motor Scythe.

GRAVELY

Gravely Overseas of Buckfastleigh in Devon introduced the American-built Gravely X Cel Estate power unit in 1950. It had a 2½ hp four stroke engine with a top speed of 3 mph, forward and reverse gears and full differential with an optional diff lock. It had a 42 in front-mounted cutter bar. A cylinder mower was among the list of power-driven attachments. The X-Cel could also be used with a plough, rotary cultivator and other implements.

A cutter bar mower was among the twenty-one attachments made for the early 1950s two wheel 5 hp Gravely Model L garden cultivator. A 43 in cutter bar was made for the Gravely Landworker in the early 1960s and twenty years later a similar cutter bar was included in the list of attachments for the 8 and 12 hp Gravely 5000 series two wheel garden tractors.

LANDMASTER

Byron Horticultural Engineering of Hucknall made a cutter bar mower in the late 1940s for the pedestrian-controlled Villiers-engined Landmaster Rotary Hoe. The 3 ft wide cutter bar, which cost £19 10s 0d, was easily attached to the front of the tractor and the knife crank mechanism was vee-belt driven from the rear power take-off shaft.

LLOYD

Lloyds of Letchworth was importing Pennsylvania hand-propelled lawn mowers from America when it introduced the Lloyd hand-propelled Autoscythe in the mid-1930s. Similar in construction to the Atcoscythe, the single wheel Lloyd Autoscythe could have either a spoked steel wheel or a pneumatic-tyred motor cycle wheel. The cutter bar on the right-hand side of the wheel was counterbalanced by the 76 cc Villiers Mar-Vil two-stroke engine and fuel tank.

The knife was driven by a vee-belt through a cam mechanism and small connecting rod to the knife head. The high-speed two-stroke Villiers Mar-Vil engine with a 2 in bore and 1½ in

5.117. Gravely made a cutter bar mower attachment for the Excel Estate power unit.

5.118. This cutter bar for the Landmaster rotary hoe was made long before the introduction of the farm safety regulations.

stroke was first made in 1934. The engine was mounted above the fuel tank but as it did not have a throttle the fuel/oil mixture was pumped directly into the crankcase.

The hand-propelled Lloyd Autoscythe was still being made in the late 1940s. The new 98 cc four-stroke JAP engine mounted on one side of the 32 in diameter single steel-wheeled scythe was counterbalanced by a 3 ft wide cutter bar with a small wheel at its outer end. An optional 2 ft wide cutter bar was available for use in confined spaces. The Autoscythe, with a spare knife, cost £65 in 1949, an optional pneumatic-tyred wheel adding £4 to the price.

The self-propelled two-wheel Lloyd Autosickle with a 3 ft wide front cutter bar cost £100 when it was introduced in 1948. Of unusual design the Autosickle had a 2½ hp JAP four-stroke engine with a Wico magneto and a Zenith carburettor. The cut grass passed between the wheels and was left in a swath behind the machine. An optional hay sweep, which cost £7 10s 0d, could be attached in front of the cutter bar.

5.119. The Loyd Motor Sickle was designed for market gardeners, poultry farmers and smallholders.

LOYD

Vivian Loyd, based at Camberley in Surrey, made the hand-propelled Loyd Motor Sickle in the 1950s. A ¾ hp Villiers two-stroke engine was used to drive the 16 in wide cutter bar, the throttle was the only control on the adjustable height handlebars and sales literature explained that no mechanical knowledge was required to maintain and operate the machine. It was also pointed out that it could be used for long periods to cut coarse grass, bracken, brambles and surplus garden produce without tiring the operator. A front rotary mower attachment was also made for the Loyd Motor Sickle.

MAYFIELD

Mayfield motor scythes were made originally at Croydon but production moved to Redhill and then Dorking in Surrey in the mid-1950s before going to Rushington near Littlehampton where the last Mayfields were built in the early 1960s. The Mayfield motor scythe cost £69 when it was introduced by Mayfield Engineering in 1949. The specification included a Villiers 1.2 hp four-stroke engine, an Albion three-speed gearbox and roller chain drive to the pneumatic-tyred wheels. The 3 ft cutter bar could be mounted centrally or offset and the knife was driven by two vee-belts and a central reciprocating mechanism. The front cutter bar was easily replaced by a toolbar with hoe blades or cultivating tines.

Mayfield Engineering made three models of the Mayfield motor scythe at Dorking for about ten years from the mid-1950s. They had 2 hp, 3 hp and 4 hp four-stroke engines with a lever-operated clutch and three-speed gearbox with an optional reverse gear. A front-mounted 3 ft wide cutter bar, similar to the Allen Scythe, was one of the fifteen different implements available for the tractor unit.

A special version of the Mayfield Motor Scythe for rough grass and brambles was introduced in 1957. It had a belt-driven vertical cutter bar at one side of the 3 ft wide cutter bar, which was easily removed when using the machine as a standard cutter bar mower. The cutter bar could also be removed from the power unit, leaving it ready for use with a toolbar and other Mayfield attachments.

John Allen & Sons, which acquired Mayfield Engineering in the mid-1960s, made Mayfield grass cutting equipment under the Allen Mayfield name until 1984. The Mayfield Eight tractor with a 3 ft or optional 4 ft cutter bar was made in the 1970s by Allen Power Equipment at Didcot in Berkshire. An 8 hp Kohler engine was used to drive the well-tested Allen-Mayfield cutter bar and propel the mower through a three-speed gearbox with an optional reverse gear. The cutter bar was driven by a vee belt with a jockey pulley clutch arrangement to a crank and knife arm. The cutter bar could be replaced with a twin-bladed rotary mower or a heavy-duty, three-blade cylinder mower in a matter of minutes.

Arun Mayfield made various models of motor scythe with 3 to 8 hp engines and a 3 or 4 ft cutter bar between 1975 and 1981. The optional cylinder mower had been discontinued but the rotary mower deck was still available.

TEAGLE

The single-wheel self propelled Jetscythe cutter bar mower with a ½ hp Teagle two stroke engine cost £50 10s when it was introduced by W T Teagle at Blackwater near Truro in 1957. There was a choice of 1 ft 6 in to 4 ft wide cutter bars with malleable iron or slightly more expensive steel fingers. Small American engines were available in the early 1960s at lower prices than the production costs of the

5.120. The four-stroke Mayfield cutter bar mower had a three-speed gearbox.

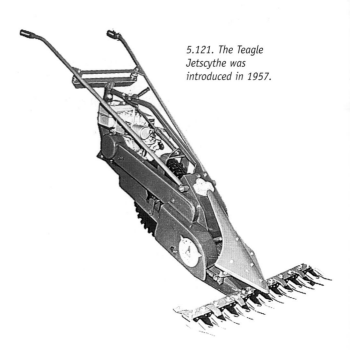

5.121. The Teagle Jetscythe was introduced in 1957.

home-made Teagle power unit so from 1965 the Jetscythe was equipped with a single-cylinder four-stroke Briggs & Stratton engine.

WOLSELEY

A 2 ft offset front-mounted Merry Tiller cutter bar was introduced in the mid-1950s. The knife crank was vee belt driven from the rotary cultivator chain case and a built-in safety slip clutch protected the drive from mechanical damage. A 3 ft cutter bar for the later and more powerful Merry Titan, which cost £150 in 1975, remained in production until the late 1970s.

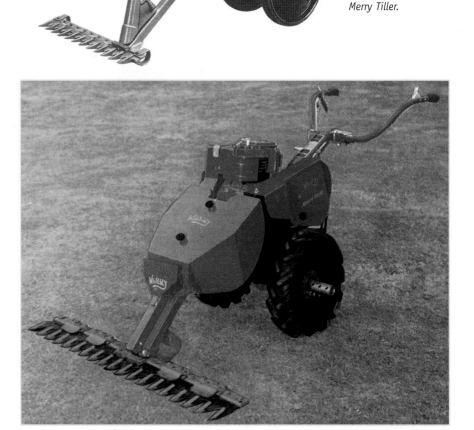

5.122. Wolseley made a cutter bar attachment for the Merry Tiller.

5.123. The Merry Sickle Mower for the Merry Tiller Titan was introduced in the late 1960s

Chapter 6

Saws, Trucks and Other Equipment

Circular Saws

Portable circular saws made in the late 1940s and 1950s for the Allen Motor Scythe, Farmers' Boy, Merry Tiller, Trusty and other two wheel garden tractors were a useful tool on smallholdings and market gardens and to a certain extent in the home garden. They were usually mounted on a separate stand attached to a garden tractor and belt-driven either from a flat belt pulley or power take-off.

The Farmers' Boy Universal saw bench was exhibited at the 1950 Royal Smithfield Show by G W Wilkin of Kingston on Thames. The 12 in diameter blade, belt driven at 2,000 to 2,500 rpm from a pulley on the tractor power take-off shaft, was capable of cutting through 5 in thick logs and planks up to 4 in wide. Although it was originally designed for the Farmers' Boy light tractor the saw bench was suitable for other garden tractors with a power take-off or pulley and it could also be driven with an electric motor. A similar saw bench with an 18 in diameter blade was made in the early 1950s for the Nash roller tractor. The bench was attached to the front of the tractor and belt-driven from an engine-mounted pulley.

The Trusty portable saw bench was a more robust piece of equipment. The optional chain-driven flat-belt pulley unit for the Trusty tractor was used to drive the saw bench and, after aligning the pulleys on the tractor and saw bench, the final task was to secure the bench in position with steel pegs knocked into the ground. Drive to the saw blade was engaged with a pair of fast and loose pulleys.

6.1. The Farmers' Boy saw bench.

6.2. The chain-driven belt pulley on the Trusty tractor could be used with a saw bench, a chaff cutter, a milking machine and other stationary equipment.

6.3. The Trusty portable saw bench was pegged to the ground to make sure it did not move, thus causing the belt to slip off the pulleys.

The Loyd Motor Sickle and the Treehog were portable circular saws with a difference. Both had a saw blade in front of a two-wheeled frame with an air-cooled petrol engine and handlebars. The Loyd Motor sickle (page 235) with a cutter bar driven by a two stroke Villiers engine was mainly used for cutting grass and scrub but the makers, Vivian Loyd & Co of Camberley in Surrey, also offered a 14 in circular saw blade at extra cost. Sales literature claimed that the saw blade could be used to fell tree saplings of up to 8 in diameter and trim protruding tree stumps down to ground level.

Advertised as a machine with a future, the Lane & Whittaker Treehog had an adjustable saw head which made it possible to cut down standing trees, cross-cut the felled timber and even trim hedges. Two models of Treehog were made in the early 1950s. The Model A with a 3 hp Villiers engine and a 27 in saw blade cost £110 in 1953.

A 5 hp Villiers engine provided the power for the 30 in diameter blade on the Model B Treehog which cost £130. The Treehog was wheeled up to a tree with the blade horizontal and, according to the manufacturers, when the tree was felled, the blade could be turned through ninety degrees and used to make vertical saw cuts to convert the trunk into logs.

THE MACHINE WITH A FUTURE

The "Treehog" One-Man Tool

for Tree-felling, Cross-cutting, Logging and Scrub Clearing.

HOG

A REAL HOG FOR WORK

LANE & WHITTAKER LTD. SHEFFIELD WORKS, HOLYHEAD RD., WELLINGTON, SHROPSHIRE, ENGLAND

Fig 6.2 A Health and Safety inspector would not be impressed by the standard of guarding on the Lane & Whittaker Treehog.

6.4. The standard of guarding on the Lane & Whittaker Tree Hog would fail to meet current farm safety regulations.

Troy Agricultural Utilities, with the impressive address of Porchester Street in London, made a front-mounted rotary mower for the 1¼ hp Troy Tractivator and added an optional 12 in circular saw to the accessory list in 1950. The Tractivator cost £41 10s 0d, the rotary mower unit was £9 17s 6d and the saw blade which replaced the mower disc was an extra £2 17s 6d. Press publicity at the time illustrated the Tractivator with the circular saw in the process of felling a tree but no explanation was offered on how the operator prevented the tree from falling on him or the Tractivator!

The Teles Smith self-propelled one-man mobile circular saw was introduced in 1949 for felling trees and cross-cutting logs. There were three models, two with a 30 in diameter saw blade and one with a 36 in blade. Power was provided by a 5½ hp JAP or a 10 hp Petter petrol engine and the saw blade, which could be set to cut in vertical, horizontal and two oblique positions, was driven by three vee-belts from a pulley on the engine countershaft.

Mayfield Engineering made a free-standing saw bench with a 10 in diameter blade; a rise and fall movement allowed the amount of blade projecting through the table to be adjusted. It was suitable for use with most 1950s Mayfield Motor Scythes, which were close-coupled to the bench. The saw was belt driven from an engine pulley.

The Mayfield logging saw, with an 18 or 24 in blade, was made in the 1950s and 1960s. It had a swing table for the timber which was pushed forwards against a belt-driven blade. The saw could be used with Mayfield and other garden tractors, a stationary engine or an electric motor. The makers suggested the Mayfield Mk 15 or another 2½ hp power tractor was required for the 18 in logging saw and the 24 in saw blade was suitable for the 4 hp Mayfield Mk 20 and similar garden tractors.

6.5. A circular saw attachment was made for the Loyd rotary sickle.

6.6. The 1950s Mayfield saw bench was used to cut timber up to 3 in thick and logs of up to 6 in diameter could be sawn with three cuts.

A 14 in saw bench for direct coupling to the Merry Tiller was made by Wolseley Engineering in the early 1960s. The saw bench had a front tubular stand and it could be used with the Merry Tiller truck at the rear for direct-loading the sawn logs. A 16 in circular saw attachment was one of several accessories made in the mid-1960s for the Atcoscythe.

Hedge Trimmers

6.7. A Tarpen hedge trimmer attachment for use with the Hayter Motor Scythe 110 volt generator attachment.

Hooks and shears were the hard way to trim hedges but help was at hand in the late 1940s with the introduction of hand-held miniature reciprocating knife trimmers for field and garden hedges. Hedge-trimming attachments were made for electric drills and for flexible drive shafts on garden cultivators and lawn mowers. Low voltage hand-held electric trimmers made for small petrol-engined generators and trimmers with a small two-stroke petrol engine helped to relieve the arm ache from long sessions with a hook or hedging shears.

A flexible drive shaft, an optional accessory for many garden tractors, was used to drive hand-held hedge trimmers made by Tarpen, Heli Strand and

other companies. Similar hedge trimmers and hedge-trimming attachments for electric drills were produced by Black & Decker, Stanley Bridges, Wolf Tools and other DIY power tool manufacturers. Small two stroke petrol engines were used to drive the Teagle Jet Cut, the Teles Clipper made by the Teles Division of E H Bentall Ltd at Maldon in Essex, the Webb Little Wonder and the more recent hand-held Mountfield hedge trimmer.

Wolseley Webb made the Webb Little Wonder hedge trimmer with a 16 or 30 in long twin reciprocating blade in the 1970s and 1980s. There were various models including 12, 110 and 240 volt electric trimmers and a Little Wonder trimmer with a small two stroke petrol engine. Mid-1970 prices started at £62 for the 110 volt model with a 16 in bar but the most expensive 30 in petrol-driven hedge cutter cost £118.

Heli-Strand Tools introduced a new 12 in hedge cutter with a flexible drive shaft in 1962. Suitable for use with most lawn mowers and garden tractors the Heli-Strand, which cost £76 10s 0d, was said to have a unique gear drive which cut growth up to ½ in thick with a clean secateur action. The hand-held Teles Tiger mains-electric hedge cutter with a 14 in cutter bar cost £18 in 1956, alternative models for use with a 12 or 110 volt electrical supply were £17 and an optional 4 ft long extension handle was available for trimming high hedges.

The Tarpen Hedgemaster introduced by Tarpen Engineering Co in the early 1950 was a hand-operated electric trimmer which could run on mains electricity or power supplied by a Tarpen generator attachment for some of the more popular makes of garden tractor. The mains-electric Hedgemaster cost £33 and a combined garden tractor generator and Hedgemaster package was £93.

The 1961 Tarpen-Strand tool catalogue included 10, 12 and 15 ft flexible drive shafts for a wide range of garden tractors from the Allen Motor Scythe to the Wolseley Merry Tiller. An independent 2¼ hp engine unit with a power take-off point and carrying handle cost £35 0s 0d and if mounted on a wheeled trolley £40 5s 0d.

The Allen Motor Scythe was one of many garden mowers and cultivators that could be used in the

6.8. A portable Tarpen-Strand engine-driven flexible shaft and 17 in hedge trimmer.

6.9. This mid-1960s Tarpen 12 in hedge trimmer was powered by a 12 volt rechargeable battery carried in a shoulder bag.

1950s with an electric or flexible shaft-driven hedge trimmer. The Allen unit consisted of a 110 volt 200 watt DC generator with 25 ft of cable and a 12 in cutter bar with a ¼ hp electric motor. Two hedge trimmers could be used at the same time with an alternative 400 watt generator. Sales literature

6.11. Hedge trimming with a Trusty tractor.

suggested that the Allen hedge trimmer could cut hedges as fast as six men using hand shears, that no particular skill was required and that its light weight made it suitable for use by either sex.

The early 1960s list of attachments for Atco cylinder mowers included a 200 watt belt-driven generator and a Tarpen hedge trimmer. Flexible drive shafts were made at the time for numerous makes of cultivator and lawn mower including the Hayter rotary scythe and Howard Bantam rotavator. A generator and an electric motor, a hydraulic motor and a small air-compressor were included in the list of accessories for the Howard Bantam.

The vee-belt driven 400 watt, 110 volt generator was mounted on a castor-wheeled trolley and had sufficient output for two Tarpen hedge trimmers or a heavy-duty Tarpen Hedgemaster. The Clipgears hedge trimmer with a 16 in cutter bar was driven by a hydraulic motor supplied with oil from a pump mounted on the Howard Bantam engine. Hedge cutting could also be done with a Shearomatic trimmer using compressed air from a small Hymatic air-compressor mounted on the Bantam Rotavator.

There were few jobs in the mid-1960s that could not be mechanised with a Wolseley Merry Tiller when as well as cutting grass and mixing concrete its flexible drive shaft could be used with a Tarpen hedge trimmer and other powered gardening aids.

6.10. An electric hedge trimmer for the Allen Motor Scythe.

The Teagle Jetcut hedge and weed cutter, introduced in 1952, cost £40 delivered to the purchaser's nearest railway station. Drive to the cutter bar was by a vee belt from a 50 cc Trojan two-stroke engine to a secondary endless roller chain drive running through a twin tubular frame carried by a strap over the operator's shoulders. The Trojan engine was found to be rather heavy and was replaced by a small 50 cc Teagle two stroke engine.

An improved version of the Jetcut, introduced in 1958, was shaft-driven through its single tubular frame. The bevel gear and drive crank mechanism was contained in an enclosed housing at the cutter bar end of the shaft. The 24 in cutter bar, said to cut through material up to ¾ in thick, could be angled through an arc of 300 degrees to cut the sides and tops of hedges or clear weeds at ground level.

The lighter Super Jetcut hedge trimmer with a 32 cc JAP engine and totally enclosed gearbox appeared in 1964. The Super Jetcut without an engine was also exported to France where it was used as an attachment for the Stihl chainsaw.

Sales literature suggested the Teagle hedge cutter would do the work of six men with hooks and that although hedge cutting could never be profitable the Teagle machine made it possible to reduce its cost. Owners of Teagle-engined Jetcuts were offered an engine replacement service priced at £4 including return postage.

Grass trimmers or strimmers with plastic cord cutters are a relatively recent introduction. Before nylon cord became readily available in the 1970s, a few companies made mechanical trimming attachments with either a rotary or an endless chain cutter.

The Stihl brush-clearing saw, introduced in the 1950s, was a saw on a stick consisting of a miniature chainsaw at one end and a small two stroke engine near the handles. Ten years later Stihl developed a rotary grass trimmer with a 10 or 12 in cutting disc driven by a 4 hp engine.

The Tarpen Grassmaster was made in the mid-1960s to cut grass and weeds in places inaccessible to a motor

6.12. Early Teagle Jetcut hedge trimmers were driven by a chain running inside the tubular handles.

6.13. Later models of the Teagle Jetcut had a shaft drive to the cutter bar.

mower or scythe. It consisted of a mains- or a battery-electric motor on a handle with a set of small rotary cutters for trimming rough grass and lawn edges. Tarpen also made a small rotary cutter head with an 8 in blade for their flexible drive shaft attachment on lawn mowers and garden cultivators.

By the mid-1980s many companies, including Allen Power Equipment, Danarm, Homelite, Husqvarna, Mountfield, Sachs Dolmar and Stihl, were marketing petrol-engined strimmers of similar design to the Danarm Whipper. Most could be used with coarse-toothed metal and tough plastic blades, a nylon cord cutter head and a saw blade. The cutter head was driven by a flexible shaft through the tubular steel handle by a small two stroke engine.

6.14. The Allen Power Equipment hedge trimmer had a two-stroke Kawasaki engine.

Power tool and lawn mower companies made small trimmers for the domestic garden in the late 1970s. Black & Decker nylon line strimmers replaced the hand shears used in many gardens to trim round trees and posts and against walls and lawn edges. Flymo introduced the Flymo Mini and dual purpose Multi Trim in 1988. Both machines had a nylon cord cutter driven by a mains-electric motor with the cord automatically fed out to the correct length when the machine was switched off. The cutter head on the Flymo Multi Trim could be adjusted to trim around trees or the edge of a lawn.

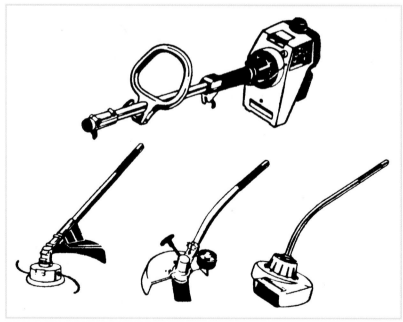

6.15. A strimmer, lawn edger and a leaf blower were some of the attachments used with the Qualcast Green Machine.

Qualcast imported the Green Machine from Long Beach, California in 1988. This was a 1 hp petrol-engined power unit on a long handle with various attachments, including a strimmer, a lawn edger, a leaf blower and a powered weeder, which were plugged into a drive shaft in the handle.

Turf Care Equipment

Lawns require well-drained soil with ample plant food to promote growth. Fertiliser was usually spread by hand and a tined fork was a cheap way of improving drainage. Various mechanical aids have been developed over the past seventy years to make these tasks easier.

Sisis Equipment of Macclesfield was established in 1932 by William Hargreaves, an experienced steam engineer. In 1962 the business moved from old buildings at Cheadle, Cheshire, to larger premises in Macclesfield.

Although Sisis had been mainly concerned with sports ground equipment, some of its products, including fertiliser spreaders and aerators, were equally useful on private lawns. Among the early Sisis implements were three- and four-pronged steel piercing forks with metal handles that cost 16s 6d and 18s 0d respectively. They were introduced in 1936; the first Sisis wheeled aerator appeared a year or two later. Scarifying the turf to remove thatch and creeping grasses is another important management task for a good lawn. The Sisis Rotorake with contra rotating tines was introduced for this purpose in 1951.

Most Sisis machines were designed for large areas of sports turf and one in particular caused great interest in the national and international press when it was announced. Although unlikely to have been used even on the largest lawns, the Sisis Drying Machine deserves description. It was taken to Headingley cricket ground in readiness to mop up water from the wicket if rain fell during the 1954 test match against Australia. Newspaper headlines enquired if it was really cricket to use the drying machine and others awaited the

umpires' decision as to whether its use would be legal.

A leaflet published by W Hargreaves & Co Ltd of Sisis Works, Cheadle quoted part of a paragraph

6.16. The Sisis Rotorake.

6.17. Sisis made a turf aerator for the Merry Tiller.

6.18. *The Sisis drying machine for cricket wickets caused quite a stir on newspaper sports pages in the summer of 1956.*

from the Laws of Cricket which stated that any roller may be used for drying the wicket and making it fit for play.

The drying machine consisted of a steel drum covered with an easily removable absorbent material and adjustable rollers which squeezed the water into tanks at the front and rear of the machine. The drum could be suitably weighted with two spring-loaded pistons to suit conditions and the water tanks were easily removed for emptying. The machine was transported on two pneumatic-tyred-wheels, which could be retracted when it was in use. Several county cricket clubs and the All England Lawn Tennis Club used the Sisis Drying machine. Indeed Essex cricket club is said to have taken one on tour with the team.

E Allman & Co of Chichester, noted for its spraying equipment, introduced a rather different-looking lawn roller in 1954. Although the Feasa looked like a roller it was in fact a fertiliser and lawn sand spreader, said to lay an even carpet of any dry material on a lawn. There were two models, one for the home gardener and a larger machine

6.19. *The Allman Feasa fertiliser and lawn sand spreader.*

for bowling greens and sports grounds. The smaller Feasa spreader cost £6 5s 6d, weighed 72 lb when filled with sand and could be adjusted to apply from 1 to 8 oz of sand per sq yd when pushed at 2 mph. The sports ground model weighed 182 lb with a full load and cost £12 15s 0d.

Simple hand-operated lawn spikers such as the hollow-tined Sheen lawn aerator, were pushed into the turf like a fork to leave cores of soil on the surface. A more expensive version which cost £9.90 in the mid-1970s had spring-loaded hollow tines. As they were pushed into the lawn plugs of soil passed upwards inside the tines to be collected in a detachable tray.

An advertisement in 1962 for the Jalo roller aerator, which cost £4 15s 0d, suggested that it made easy work of aerating a lawn with its steel spikes making 50 holes per second when pushed over the turf.

The Tudor aerator was an example of a roller aerator with a number of tine bars between two wheels making holes about 1½ in deep as it was

6.21. Allen garden sweepers were first made in 1972.

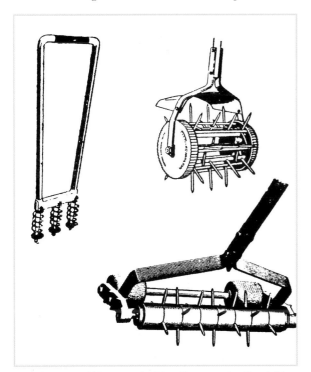

6.20. Lawns could be aerated with a Sheen aerator (top left), a Tudor spiker (top right) or a Jalo roller spiker.

6.22. The Ransomes leaf blower put the leaves in windrows for collection.

6.23. Iseki made a rotary brush for some of its garden tractors.

pushed across a lawn. The 9 in wide Tudor aerator cost £6.75 in 1973 and the 12 in machine £7.50.

The autumn chore of sweeping up leaves is another task to have attracted the attention of the inventive mind. Hand-propelled sweepers made by Allen, Atco, Wolf and other companies with a brush to throw the leaves into a suitable container hooked on the handlebars have been available for many years. Much wider leaf sweepers, notably those made by Allen in the 1970s, were self propelled or towed by a small tractor. A 2½ hp Briggs & Stratton engine was used to propel the 30 in wide pedestrian-controlled Allen Scorpio sweeper, which cost £120 in the mid-1970s.

A rotary brush provided another method of sweeping leaves into a swath for collection and the same result could also be achieved with a leaf blower. The Ransomes Leafblower with a fan driven by an 8 hp Briggs & Stratton engine was one such

machine. Sales literature for the mid-1980s Ransomes Leafblower indicated that the ¼ in thick fan blades would clear leaves from walkways and leave them in rows for collection.

Atco came up with an alternative method of tidying the garden with the introduction of the Blow & Vac in 1987. A 31 cc engine and fan unit was used to suck leaves and debris into a large capacity bag carried on the shoulder or to blow them into convenient piles for hand collection. The Blow & Vac was discontinued in 1989.

Flymo launched the Garden Vac, described as a portable garden vacuum cleaner, in 1993. A 650 watt mains-electric motor was used to drive the fan, which could be set to blow leaves into piles and then switched to the vacuum position to suck them into a bag on the machine. The vacuum mode had two settings, for light and heavy garden debris.

Manure, compost, soil, garden rubbish and even

Trailers and Trucks

firewood were and still are moved around a garden in a wheelbarrow. However, smallholders and market gardeners, who were disinclined to push a barrow, could invest in a motor truck or a trailer for their garden cultivator. Gardeners, nurserymen and smallholders used several types of motor truck including the Batric Truc Tractor, the Geest Tug, the Martin, the Trac Grip, the Tubo Truc Tracta and the Wrigley made mainly during the 1940s and 1950s. These useful little vehicles served their purpose for many years until similar machines including the tracked Kubota transporter replaced them in the mid-1990s.

The lightweight three wheeled Wrigley motor truck with a Villiers engine mounted on a turntable above the single front wheel was invented by Arthur Wrigley and his brother in the

6.24. Whitlock Brothers made trailers for Ransomes MG tractors.

6.25. The first Wrigley motor trucks were made in the early 1930s.

early 1930s. They founded Wessex Industries at Poole in Dorset a year or so later to manufacture the Wrigley motor truck.

Wessex Industries, which made fixed and screw-tipping farm and horticultural trailers, introduced a 5 cwt capacity low-loader Wrigley motor truck in 1948. Powered by a 1 hp 98 cc Villiers Mk 10 four stroke side-valve petrol engine, the Wrigley truck had a three-speed Albion gearbox with a built-in clutch and roller chain drive to the front wheel. The truck was steered with a hinged tubular-steel handlebar swinging backward or forward over the engine and fixed to the 14 in turntable revolving on ½ in diameter steel balls.

The turntable carried the complete engine, transmission, controls and single front wheel which could be turned through 360 degrees so that the user could steer the truck while riding on it or lead it like a horse from the front. The rear wheels had expanding shoe brakes operated with a foot pedal and there was a hand brake on the front wheel.

A larger Wrigley truck, added in 1949, had a 3 hp 250 cc side-valve Villiers engine, an Albion or Burman gearbox with three forward speeds and reverse and a top road speed of 10 mph. It was designed to carry a maximum load of up to 1 ton on level ground but the capacity was reduced in less favourable situations. Various truck bodies were made, including a hydraulic tipper with a hand-operated pump.

In the mid-1950s Wrigley three-wheel motor trucks included a 10 cwt capacity model with a Villiers engine, a three forward speed gearbox and roller chain drive to the front wheel, all mounted on a turntable with the same 360 degree steering angle. The larger 20 cwt capacity Wrigley truck had a 3 hp Villiers engine. There were several body options, including a flat platform with fixed, detachable or hinged sides and tailgate, while

another model had an extended chassis with a fixed seat and foot board. A narrow industrial truck was made for use in gangways and the stillage model had a hydraulic platform with a hand-operated pump.

Forklift manufacturers Montgomerie Reid of Basingstoke bought Wessex Industries when the company ceased trading in the late 1960s. The motor truck side business was then sold to Wantage Engineering in 1974 when the Wrigley name was revived and production of both ride on and pedestrian-controlled ½ ton and 1 ton trucks began at Wantage. Factories, warehouses and horticultural enterprises were the main customers for the truck, which had hardly changed from the original design, but there was a choice of body styles with petrol, diesel or electric power units.

Either a Honda petrol engine, which could be converted to run on LPG, or a 7 hp air-cooled Lister diesel engine provided the power for the early 1980s Wrigley trucks. They had a three forward and one reverse gearbox with a top laden speed of 6 mph. They were still handlebar-steered but an upholstered driving seat provided far more comfort than in

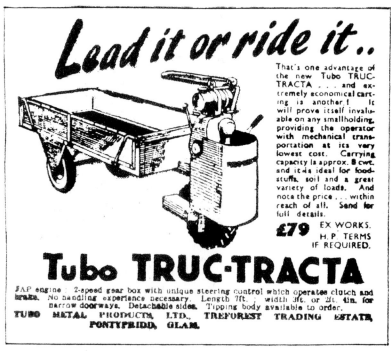

6.26. The Tubo Truc-Tracta.

earlier days when the driver sat on a short wooden plank or a wooden box. In the mid-1990s Wantage Engineering manufactured improved versions of the three-wheel Wrigley Truck, still in a style similar to those made by Arthur Wrigley sixty years earlier.

The first Lister Auto-Trucks, designed by the Auto-Mower Engineering Co near Bath, had an air-cooled JAP engine and were made in 1926. There were various body options including fixed and tipping load platforms, and several thousand were sold by the mid-1930s. The much later Lister LD Auto-Truck with a 3½ hp Lister air-cooled diesel engine and transmission unit mounted on a single turntable-steered front wheel cost £198 when it was introduced at the 1956 Smithfield Show. The LD truck with a maximum payload of 2,000 lb was claimed to cost no more than 3d an hour to run. Lister Auto Trucks were still being made in the late 1990s.

The first Tubo Truc Tractas were made in 1951 by Tubo Metal Products at Pontypridd in Glamorgan. Like the Wrigley truck, the user could either lead it or ride on the seat. It was advertised in 1951 at £79 ex works and extremely economical carting was claimed to be one of its advantages. Power to haul a load of up to 8 cwt was provided by a 1¼ hp JAP four stroke engine through a two-speed gearbox to the front wheel. The steering handle also engaged the clutch and front wheel brake.

A wider range of Truc Tractas was made in 1954 with prices starting at £35 for the Junior model, £65 for the standard machine and £89 10s 0d for the 10 cwt capacity Estate truck. The earlier Villiers or JAP engines were replaced with a 2 bhp BSA engine and the standard Truc Tracta had a three forward and one reverse Albion gearbox with the option of 5 or 8 mph top speed and roller chain drive to the front wheel.

In common with the 8 cwt model the steering handle on the Estate version operated the clutch but the expanding shoe brakes were pedal operated and a hand brake was provided for parking. The Junior Truc Tracta with a 1¼ hp two stroke JAP 80 engine was said to be the cheapest motor truck on the market at the time. The 4 ft by 2 ft 6 in load platform carried up to 5 cwt at a maximum speed of 3 mph and a tipping platform was available at extra cost.

Hatch Brothers of Callington in Cornwall made three-wheeled Trac Grip motor trucks in the 1950s. Improved versions were introduced from time to time, including a new 10 to 12 cwt capacity model in 1954. Unlike some of the earlier trucks with the single wheel at the rear, the new model had a single front wheel and was driven by a 2 hp or optional 3½ hp BSA engine through a differential rear axle to the rear wheels.

The 1954 Trac Grip tipper truck, which cost £170, had a cushioned seat for the driver and, unlike most other trucks at the time, a steering wheel instead of the more usual handlebars with a chain and sprocket linkage to the front wheel.

The 10 cwt capacity Trac Grip power barrow with a 3 hp engine, a two forward and one reverse gearbox and single driving wheel on a front turntable was introduced in 1964. It was handlebar steered and there was a choice of a metal dumper body or three sizes of flat deck body narrow enough to pass through a 3 ft 6 in doorway. Manufacturing rights for the Trac Grip trucks were acquired at a later date by Wagstaff and Gladwell Ltd who marketed them as the WAGE truck with blue and silver paintwork.

The 1950s Geest Tug was, as its name suggests, a towing unit rather than a truck but it did have a small load-carrying platform behind the driving seat. Made by Geest Industries at Spalding, Lincolnshire, it had a Villiers air-cooled engine and an Albion three forward and one reverse gearbox mounted on a turntable above the single front wheel.

Geest also made two models of motor truck in the early 1950s. Both models had a Villiers air-cooled engine and a three-speed Albion gearbox mounted on a turntable over the single front wheel. The 1¼ hp MT 2/1 truck with top speeds of 1, 3 and 6 mph had a 10 cwt load capacity while the 3 hp one ton MT 2/3 truck had top speeds of 2, 5 and 12 mph. The trucks were handlebar-steered from a seat at the front of the load platform and there was a choice of an industrial- or agricultural-type tyre for the front wheel.

The Batric Tugtractor made at Stroud in the early 1970s was a three wheeled electric motor truck that could travel six miles with a fully charged 12 volt battery. Adding one or two extra batteries extended its range to either twelve or eighteen miles recharging was necessary. The Tugtractor was not a tug but resembled a pedestrian-controlled handcart

6.27. The Bonser truck cost £216 in the early 1960s, the optional front wheel bumper bar adding £4 5s 0d to the price.

with the motor and batteries under the load platform. Early models had a 12 cu ft capacity dropside or tipper body, but an improved model with a lower chassis was introduced in 1974.

The Tugtractor on pneumatic tyres, with a forward and reverse control unit, cost about £200. An infinitely variable electronic speed control system was available at extra cost. The truck was steered with a tubular handle linked to the single rear castor wheel and by using either forward or reverse gear the operator could either walk in front of or behind the truck. A socket outlet on the Tugtractor provided a power source for auxiliary lighting and power tools including a battery-operated hedge trimmer. Attachments included a front-mounted scraper blade and a detachable tipping bin for carrying fertilisers, manure, etc.

More powerful trucks including the Bonser and Opperman Motocart carried much heavier loads. The 5 hp Bonser three-wheel tipper truck made by Bonser Equipment Ltd at Hucknall near Nottingham had a 13 ft 6 in turning circle and when fully laden it carried one ton. The specification included three forward speeds and one reverse, rear wheel drive through a

6.27a The engine and transmission were mounted on the front wheels of the Opperman Motocart.

hypoid differential axle and 9 in Girling shoe brakes.

The Opperman Motocart (page 129) with the choice of a wooden 30 cwt fixed or tipper body and a single-cylinder 8 hp air-cooled engine at one side of its large diameter front wheel was made by S E Opperman Ltd of Boreham Wood from the mid-1940s.

The Martin 1 ton truck with a hand- or hydraulically tipped body and a 7 or 10 hp Kohler engine was made in the mid-1970s by Malcolm Martin Trucks of Kirby in Ashfield, Nottinghamshire. A shortened version of the 12 hp tricycle-wheeled motor truck with a rear drawbar was used to tow a trailer and other equipment.

Although small dumpers were intended for building work, some market gardeners used one to move goods and material around the holding. The 1950s Nash Roller Tractor (page 126) was a dumper with a difference in that with the rear wheels removed it could be used as a land or road roller. A trailed tipping dumper truck was made for the Barford Atom garden tractor in the early 1950s. The 3 cwt capacity two-wheeled truck could be attached to the front of the Barford Atom or hitched to the tractor drawbar.

Tractors (London) Ltd, well known for Trusty two- and four-wheeled tractors, added dumper trucks and an angle dozer to its product range in 1957. The two dumper trucks with many parts identical to those used for the Trusty Steed tractor, had skip capacities of ⅓ and 1 cubic yard. There was a choice of a single-cylinder 6 hp air-cooled JAP or a Petter

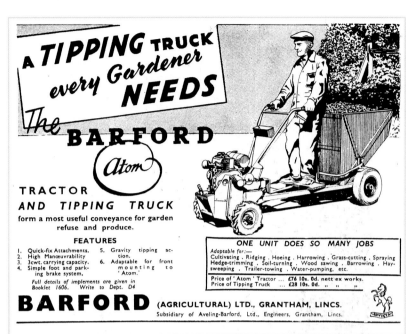

6.28. The Barford Atom tipper truck cost £28 10s 0d in 1951.

AVA 1 diesel engine for both models which had a three forward and reverse gearbox, independent wheel clutches for sharp turns and a drawbar for towing other equipment.

6.29. There were two sizes of the Trusty Dumper truck.

6.29a The Autobarrow 1 ton truck was made at Evesham in the mid-1970s. It had an 8 hp Kohler petrol engine, Albion four forward and reverse gearbox, electrically operated tipping ram and an African hardwood body on a steel frame.

6.29b The self-propelled mid-1970s Bentall Loadadump with a tipping hopper had a 3½ hp four-stroke Aspera engine, dog clutch and a single-speed gearbox.

Co also made the small self propelled Catterbug tracked load carrier at the time. The Catterbug had a 2 cwt payload and was driven by a 171 cc four stroke engine with a twist grip throttle and an automatic clutch. Drive was by vee belt from the engine pulley to a large diameter pulley on the drive shaft carrying its wide slatted track gear.

The early 1990s pedestrian-controlled Kubota tracked carrier was a very sophisticated machine compared with the Catterbug. It had a 5.5 hp Kubota petrol engine, two forward and one reverse gears and could carry about 6 cwt.

Speed and comfort were added attractions of motorised trucks by the mid-1980s. The John Deere AMT with high flotation tyres, four wheel drive and steered by a fifth wheel at the front could go virtually anywhere. It had a variable speed belt-driven transmission, optional hydraulically tipped body and seats for driver and passenger.

Two wheeled garden tractors of all sizes were used to move materials around with a trailer in the 1940s and 1950s. A trailer was among the attachments used with the Rowtrac in the mid-1930s and Barford (Agricultural) Ltd made a 7½ cwt trailer with internal expanding shoe brakes in the early 1950s for the Barford Atom 15 garden tractor. It had a wooden platform seat for the driver and the Barford Atom catalogue pointed out that the side and end boards could be removed when necessary.

Wolseley Engineering added the Merry Truck to a long list of equipment for the Merry Tiller in 1960. Most garden cultivator trucks were towed behind the tractor with a seat for the person steering the outfit. The Merry Truck was different as it had a circular frame attached to the Merry Tiller secured with a single pin. Once in position, the Merry Tiller could be turned through a full circle within the frame so that it could either pull or push the truck. The operator had to walk behind and steer the Truck with the Merry

The Moto-Bara made by Hylett Adams at Stonehouse near Gloucester in the mid-1960s was a pedestrian-controlled motorised handcart with its cushion-tyred wheels belt-driven by a 2½ hp Briggs & Stratton engine. It was handlebar-steered from the rear and the 5 cwt capacity hopper body was tipped forward to empty its contents. A flat truck body was available as an optional extra.

The hover truck principle was used by the Light Hovercraft Co of East Grinstead in the mid-1970s for the Hover Pallet, which was said to be able to traverse almost any terrain on a cushion of air. It had a load platform for sacks, bales, etc and depending on the model could carry up to 10 cwt. The Light Hovercraft

6.30. John Deere AMTs (all-materials transporter) were being made in the mid-1980s.

Tiller handlebars when it was pushed but he or she could walk alongside or ride in the trailer when it was towed. The Merry Truck (page 294) carried up to 5 cwt and the body could be tipped to empty out bulk loads.

A carrier attachment first made for the Allen Motor Scythe in 1949 (page 295) also served as a platform for a chemical tank when spraying fruit trees with a hand lance. The load platform was secured above the wheels with four bolts and a small wheel mounted on the cutter bar bracket supported the front end. With the front board removed the carrier could be tipped to empty out bulk materials.

6.31. The sack on this mid-1930s Rowtrac trailer was a convenient driving seat.

6.32. Merry Tiller attachments included a Baromix concrete mixer (A), the 5 cwt capacity Merry Truck (B) and a flexible drive shaft for a hedge trimmer (C), a tool chuck and a cattle clipper.

6.33. The Allen Motor Scythe load carrier.

6.34. This two-wheel trailer was specially made for the Gunsmith light tractor.

Trailers for small garden tractors held little more than a barrow load but larger models were made to match the increased size of more powerful three- and four-wheeled ride-on tractors such as the Gunsmith, the Garner and the Trusty Steed. This trend continued with the arrival of compact tractors and even larger trailers were made for them. However, as is so often the case, the need for small tractors and trailers for large gardens was ignored in the rush to make bigger and better equipment. Eventually a new generation of garden tractors and ride-on mowers equipped with a rear drawbar arrived in the 1970s and trailers were made to match.

6.35. Reinforced sides, corners and floor were features of the Trusty trailer.

6.36. Massey Ferguson included a trailer in the range of equipment for the MF 1010 compact tractor.

Index

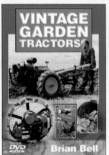

Fifty Years of Farm Tractors

Hardback book by Brian Bell

Dealing with over a hundred companies, in an A-Z format, Brian Bell describes the models and innovations that each contributed to post-war tractor development. With over 300 illustrations, the book is a unique guide to the wide range of machines to be found on Britain's farms.

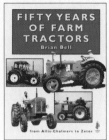

Farm Machinery Fifth Edition

Hardback book by Brian Bell

A key text for agricultural students, Brian Bell's book includes chapters on tractors, cultivation and drilling machinery, also crop treatment and harvest machinery. Further sections deal with farmyard and estate maintenance machinery, mechanical handlers, dairy equipment, irrigation, farm power and the farm workshop. Safety is stressed throughout.

Tractor Ploughing Manual

Hardback book by Brian Bell

This Society of Ploughmen handbook introduces the types of plough and how to set them up. Then it covers in detail the main classes of competitive tractor ploughing: vintage mounted and trailed; world-type conventional ploughing; and reversible ploughing. Highly illustrated.

Vintage Match Ploughing Skills

DVD by Brian Bell

This exceptional programme shows the techniques of both tractor-trailed and mounted match ploughing. Featuring two champion ploughmen – Ken Chappell and Michael Watkins – the programme carefully takes the viewer through demonstration plots showing the opening, ploughing out and finish, and in particular gives clear guidance on how to identify and correct common faults.

Ploughs and Ploughing Techniques

DVD by Brian Bell

This instructive programme is a high-quality guide to commercial reversible ploughing. It begins with an appraisal of the types and makes of multi-furrow ploughs and then shows in detail how to prepare and adjust a plough for work It concludes with headland marking, field procedures and ploughing techniques.

Brian Bell's Classics Series *DVDs*

Classic Tractors

Featuring archive film footage, much from manufacturers' promotional films, this programme contains a wealth of popular and rare tractors working in a variety of situations. The emphasis is on 1945 to the 1990s.

Classic Combines

Beginning when the binder reigned supreme, this programme includes rare footage of early harvesting machines, including those manufactured by IH, Ransomes, Fisher Humphries and Claas. The DVD brings the combine story up to date with extracts from films by Claas as well as John Deere, New Holland and MF.

Classic Farm Machinery 1

A collection of archive farm machinery films, covering the period 1940-70. Most of the well-known manufacturers are represented, as well as some less familiar names. A broad range of equipment has been chosen to demonstrate developments in machinery for ploughing, cultivating, drilling, spraying and harvesting.

Classic Farm Machinery 2

A programme showing advances in farm mechanisation from the 1970s to the 1990s, using archive footage from most of the major manufacturers. Dealing with equipment for ploughing, cultivating, drilling, spraying and harvesting, the video highlights the emergence of ever more powerful machines with electronic controls.

About the Author

Brian Bell MBE

A Norfolk farmer's son, Brian played a key role in developing agricultural education in Suffolk from the 1950s onwards. For many years he was vice-principal of the Otley Agricultural College having previously headed the agricultural engineering section. He established the annual 'Power in Action' demonstrations in which the latest farm machinery is put through its paces and he campaigned vigorously for improved farm safety, serving for many years on the Suffolk Farm Safety Committee. He is secretary of the Suffolk Farm Machinery Club. In 1993 he retired from Otley College and was created a Member of the Order of the British Empire for his services to agriculture. He is past secretary and chairman of the East Anglian branch of the Institution of Agricultural Engineers.

Brian's writing career began in 1963 with the publication of *Farm Machinery* in Cassell's 'Farm Books' series. In 1979 Farming Press published a new *Farm Machinery*, which is now in its fifth enlarged edition, with more than 25,000 copies sold. Brian's involvement with videos began in 1995 when he compiled and scripted *Classic Farm Machinery Vol 1*.

Brian Bell writes on machinery past and present for several specialist magazines. He lives in Suffolk with his wife Ivy. They have three sons.

Books and DVDs by Brian Bell

Books
Farm Machinery 5th Edition
Fifty Years of Farm Machinery
Fifty Years of Farm Tractors
Machinery for Horticulture (with Stewart Cousins)
Ransomes, Sims and Jefferies
Seventy Years of Garden Machinery
Tractor Ploughing Manual

Videos / DVDs
Acres of Change
Classic Combines
Classic Farm Machinery Vol. 1 1940-1970
Classic Farm Machinery Vol. 2 1970-1995
Classic Tractors
Farm Machinery Film Records Vol 1 Grain, Grass and Silage
Farm Machinery Film Records Vol 2 Autumn Work and Rootcrops
Harvest from Sickle to Satellite
Ploughs and Ploughing Techniques
Power of the Past
Reversible and Conventional Match Ploughing Skills
Steam at Strumpshaw
Thatcher's Harvest
Tracks Across the Field
Vintage Match Ploughing Skills
Vintage Garden Tractors